On The Right Lines?

On The Right Lines?

The Limits of Technological Innovation

STEPHEN POTTER

St. Martin's Press, New York

First published in the United States of America in 1987

Printed in Great Britain

Library of Congress Cataloging-in-Publication Data

Potter, Stephen, 1953–
On the right lines?

Bibliography: p.
Includes index.
1. High speed ground transportation. I. Title.
TF 1450.P68 1987 385 86-27894
ISBN 0-312-00488-5

Contents

List of Figures and Tables

Figures

Tables

Acknowledgements

I would like to thank the following people for information, advice and comments during the preparation of this book and the Open University case study that preceded it:

Dr Robin Roy, Senior Lecturer in Design, the Open University, with whom I worked on the original Open University T362 case study and some of whose material has been included in this book.

John Fogg, Dr John Prideaux, Dr Richard J. Todd and Dr Bernard Nield, British Railways Board.

Dr Alan Wickens and Dr David Boocock, Mr David Rollin, Railway Technical Centre, Derby.

J. K. Steer, Steer, Davies and Gleave Ltd.

Dorothy Griffiths, Imperial College

Colin Ledsome, The Design Council.

Professor Sir Hugh Ford, Ford and Dain Research Ltd.

Roger Ford and James Abbott, *Modern Railways Magazine*.

British Rail Photographic Library.

GEC Transportation Projects.

Modern Railways and New South Wales State Rail Authority.

Travel Lens Photographic, Leeds

1 Why so many 'might have beens'?

This book seeks to explore the question of why, while there have been many proposals for technologically advanced ground transport methods, virtually none of these have got very far in practical commercial terms. We live in an age which is almost obsessed by technological advance; where thinking through the economic and social consequences of such technological advances lags so frighteningly far behind the technology itself. Yet there are some areas emerging where it looks as if 'advanced' technology has virtually been bypassed or ignored. Could it possibly be that technological advance is nearing a limit?

Such a concept seems almost alien to prevailing thought today, but forces do exist that set boundaries to innovation. This book is a case study that explores such forces. The case study chosen is that of high-speed passenger ground transport. In the main this means fast trains, but the reason why this is so is surely part of the story.

Perhaps it would be most appropriate to begin by setting you the reader, a little puzzle. For this to be successful, I trust that you have been terribly conventional and have started to read this book at the beginning. In so doing you will not have already discovered the answer to this puzzle, which appears on the next page. The puzzle is simply this. Over the page is a photograph (Figure 1) of a very novel and innovative train. It is a monorail, but it *stands* on an ordinary single railway line. The first part of the puzzle is simply the question 'why doesn't this train fall over?'

The train was the Brennan monorail and the reason why it was able to stand on a single rail was that it was

Figure 1. The Brennan monorail

stabilized by contra-rotating gyroscopes set in a vacuum chamber. This gave the train a very smooth ride, it could bank around steep curves (almost like a bicycle!) and had the potential for very high speeds. This prototype train was built after a series of experiments with scale models had shown the concept of a gyroscopically stabilized monorail to be technically sound. This railcar could carry thirty passengers along a specially constructed test track.

Should the train suffer from engine failure, the gyroscopes, being set in a vacuum on low-friction bearings, would continue to rotate and keep the train upright long enough to position supports beneath it.

The technology is clearly fascinating and spectacular. It has all the makings of a good popular TV science programme. But I am speaking of this train in the past tense, for although this technically successful prototype was built, the Brennan monorail never saw commercial service.

The second part of the puzzle is simply guessing the date of this photograph. When would you say this project dated from? Would it be:

(a) 1900–1914
(b) 1914–1925
(c) 1925–1939
(d) 1939–1950
(e) 1950–1960
(f) 1960–1970
(g) after 1970

The answer is at the foot of the next page

I found the account of the Brennan monorail in an encyclopaedia together with plans for other, somewhat more conventional, high-speed monorails, steam turbine trains and extremely novel urban mass transit systems. All would have made fascinating TV science programmes today, but at the time of their development the invention of the television was still some twenty years in the future.

Indeed, the juxtaposition of these fast ground transport projects with other entries in the encyclopaedia struck me as almost ludicrous. The section on aviation consisted of fragile string and canvas byplanes (the Wright Brothers' pioneering flight having taken place only eight years before). An article on recording showed how acoustic records were cut in wax using vast horns in a necessarily cramped recording studio. The contrast in technological progress between these and the railway section of the encyclopaedia could not have been stronger.

This century has seen plans for technically advanced rail (and other ground transport methods) come and go while the classical railway has gradually evolved and improved. Railways may well have come under threat from vast technical developments in air and road transport, but they have by no means suffered anything like the fate of the more technically advanced rail alternatives. A variety of monorails have given way to projects such as the tracked hovercraft and magnetic levitation. Yet in practice, although such systems have had some small-scale application, quite modest advances in conventional trains have totally dominated this sector. A handful of monorails operate around the world and one magnetic levitation train runs for a few hundred metres between Birmingham's airport and its station.

To say that technical innovation has reached a limit would be too simplistic a conclusion to such evidence. Rather, we have in the development of advanced fast passenger trains an excellent case study of the forces that mould, direct, constrain and (in many cases) stimulate innovation. These forces can be categorized in many ways, but may roughly be grouped as economic, historical, socio-political and managerial. This book examines these forces, focusing on Britain where they seem to have been strongest. The blend of these forces seems to vary significantly between countries, which helps to explain different national stances to fast rail developments.

It cannot be denied that a study of fast passenger trains has a fascinating attraction in its own right (as any train enthusiast would tell you!). This attraction is not denied, but this book is primarily seeking to identify lessons from this story which show how the direction and scope of technological innovation may be determined. We begin by returning to the Brennan monorail, to examine the wider context in which this promising, technologically advanced project got absolutely nowhere.

2 The development of fast trains, 1930–1970

What has been will be again,
What has been done will be done again.
There is nothing new under the Sun.

King Solomon
(Ecclesiastes 1.9)

Before considering the forces that constrain and stimulate the development of modern high-speed passenger trains, it is worthwhile delving into a little railway history. This historical legacy itself, in terms of technology, infrastructure, economics, management and politics, has had a major influence on the options available to different railway systems for high-speed developments. It also illustrates well forces that are still at work today.

The development of fast trains in Britain

The development of fast passenger trains in Britain cannot be viewed outside the context of the development and organization of Britain's rail network as a whole. The railway network was virtually all built in the nineteenth century under technical and economic conditions very different from those prevailing today. The railways dominated transport and they invested heavily in technology and organizational structures that were appropriate to their era: metal tracks (quite an innovation at the time!); the steam locomotive; and signal, control and operational methods geared to the availability of cheap, relatively unskilled, but highly disciplined labour. Because of the sheer scale and long life of this nineteenth-century investment, its historical legacy has a significant impact on fast ground transport developments today.

Indeed, even at the beginning of this century there existed potentially radical rail innovations that made no headway. The introduction to this book briefly examined the Brennan monorail, one of several undeveloped technologically advanced rail projects of the Edwardian era. The original stimulus for its development was the rising cost of new railway infrastructure, which had become a major barrier to the further expansion of Britain's rail network. However, it was soon realized that such a train had a potential for very high speeds—200 km/h (125 m.p.h.) or more. But, as a monorail, it could not run on existing track, and by 1910 there were more than enough main lines already built. The technology was really too advanced and too expensive for branch lines, where economies of construction could be achieved by much less ambitious means, such as the use of diesel railcars or light narrow-gauge lines. For fast running over trunk routes, the monorail would have required the rebuilding of all existing track, together with the realignment of curves to allow for its banking angle. There was no economic case for such a vast investment. Interest in the Brennan Monorail faded and today it is little more than a fascinating 'might have been'.

Because the railways are such a large industry, with a vast amount of investment in assets that last a long time (the lifetime of a modern train is usually thirty years and can be more), rail developments throughout the world have been characteristically 'evolutionary', involving the improvement of known designs rather than major technological innovations. Indeed, the definition of the term 'innovation' is quite difficult within this context, for rarely has there been a totally new idea or application in the rail industry. Most innovations have emerged gradually as earlier practices have been modified. This process has become self-reinforcing, as engineering departments and management have organized themselves around such an evolutionary approach. Well might King Solomon's words have been a prophecy on the innovation process in the railways!

High-speed steam: an engineering irrelevance?

The design and development of Britain's first commercial high-speed train in the 1930s illustrates this evolutionary tradition. When talking about 'high-speed' or 'high-performance' trains it is difficult to define exactly what these are since the borderline between 'high-speed' and other trains can be indistinct. Around the turn of the century many competing railway companies sought to better their rivals in terms of speed. These railway races were largely unofficial, as was the case on the Plymouth–London route between the Great Western Railway and the London and South Western Railway. This was a prestige route, involving mail and passengers from transatlantic liners docking at Plymouth. In May 1904, on an official record-breaking run, one GWR locomotive (the 'City of Truro'), was timed at 164.6 km/h (102.3 m.p.h.). This official publicity stunt led to unofficial emulation on a number of occasions. Freeman Allen (1978, p. 14) aptly catches the mood of the time:

> Knowledgeable Americans off the transatlantic liners, it was said, were frequently passing a fistful of dollars on to the LSWR drivers in hopes of a record sprint to London, and the enginemen were giving the visitors value by such hectic excesses as hitting the sharp curves through Salisbury at twice the stipulated speed limit.

After a few months the LSWR clamped down on the races and drivers were heavily disciplined for exceeding track speeds. But, on 1 July 1906, one of the LSWR boat trains entered the tight Salisbury station curve, on which there was a 30 m.p.h. limit, at around 70 m.p.h. Lurching over, the train caught an empty milk wagon, and ploughed into a goods engine. The train crew of four and twenty-four passengers were killed. Probably unjustifiably, the newspapers blamed the crash on railway racing. Whatever the cause, the Salisbury accident gave high-speed operations a bad name. It is little wonder that the Brennan monorail received scanty interest from the railways as a high-speed project.

At much the same time, a similar railway race was taking place in the United States. What became known as the 'Great Speed War' broke out between the Pennsylvania Railroad and the New York Central for the long-haul New York–Chicago passenger business. In 1900 this journey took about twenty-eight hours, which was gradually cut, by one company or the other, to eighteen hours by 1905. Technologically, the Pennsylvania had the winning card in their powerful E2 Atlantic locomotive. The inaugural 'Pennsylvania Special' was recorded as averaging 110.3 km/h (68.6 m.p.h.) over one section of the route, although the top speed claimed of 204.5 km/h (127.1 m.p.h.) is now viewed as highly unlikely. But, despite the excitement generated by this, the battle for passengers was actually won by the New York Central. Their salesmanship and marketing won more customers than the Pennsylvania's speed—a salient lesson which, like many in the railway industry, was destined to be repeated in later eras.

The return of high speed

To return to Britain, by the 1930s the railways companies were beginning to face competition other than from each other. The airlines were beginning to become serious competitors, and, among the better-off, the car was becoming the fashionable alternative to the train. It was under these circumstances that speed returned to the railways' agenda in the form of Britain's first purpose-built high-speed train.

Although a specified top speed usually formed part of the design brief of a locomotive, Sir Nigel Gresley's A4 Pacific was the first train built in Britain for which speed was the primary design objective. Ordered by the London and North Eastern Railway (LNER), the first of these powerful streamlined steam locomotives (the 'Silver Link') entered public service in September 1935. For the London to Newcastle route the schedules required an average speed of 112.6 km/h (70 m.p.h.), involving sustained running at up to 160 km/h (100 m.p.h.). In fact, in a trial run a few days before entering public service, the 'Silver Link' reached 181 km/h (112.5 m.p.h.) and, although generally limited in public service to 160 km/h (100 m.p.h.), another

Figure 2. An A4 Pacific-hauled train
Note the streamlined, wedge-shaped nose, derived from wind tunnel tests

A4 (the 'Silver Fox') reached 181.8 km/h (113 m.p.h.) in passenger carrying service in August 1936. In July 1938, the A4 'Mallard' set the world's absolute speed record for steam traction at 202.7 km/h (126 m.p.h.).

The design of this train involved some of the most sophisticated techniques available at the time. For example, both the LNER and the London Midland and Scottish Railway (LMS) used the wind tunnel facilities at the National Physical Laboratory in a programme of joint research on locomotive design. The LMS then decided to invest in their own wind tunnel at Derby, which was used by both companies. This research indicated both the power output required to overcome wind resistance on a train at high speeds and the consequent implications for train design.

In the 112–160 km/h (70–100 m.p.h.) range it was discovered that air resistance absorbed 85 per cent of traction power. Streamlining thus became a vital feature of the A4 Pacific, with its distinctive 'wedge' shaped nose, plus a host of streamlining features to the carriages so as to reduce drag-inducing eddies. These included a streamlined rear carriage, dipped side panels to the carriages and rubber sheeting to close the connecting gaps between the carriages, so improving air-flow at the sides. Weight-saving was also understood to be important and this train had carriages with articulated bogies, i.e. the wheel bogies were shared between some of the carriages, reducing the number needed. This feature was to return with the Advanced Passenger Train in the 1970s.

The control of innovation

But the use of 1930s high technology was entirely *within* the context of a steam-hauled train on steel tracks, controlled by labour-intensive signalling equipment and fitting in with the timetabling and service operations of a normal railway. The innovation of a streamlined fast steam locomotive and specially designed coaches was but a small part of a large and complex operation, which could only gradually alter as each component was updated and amended. The innovative process for one element could not go beyond the limitations imposed by the whole. For example, the operation of an A4 Pacific required the clearing of the track ahead of any slower trains. Given that normal passenger and freight trains at that time operated at around 60–90 km/h this meant no departures on a particular line for an hour or so beforehand! As such the operation of a single fast train imposed great strains on the timetabling and general operations of the railway.

The evolutionary nature of rail innovations reflects the nature of the rail industry itself, but there were aspects in the development of fast steam trains that require further explanation. Why, for example, was there such a strong commitment by Britain's leading rail engineers to steam power when almost every other country was moving towards diesel or electric traction? The breakthrough in diesel traction had come in 1930, when the United States company, Electromotive, developed the revolutionary 201A engine. This quadrupled the power/weight ratio compared to previous diesel engines. This was the engine installed in the Budd 'Pioneer Zephyr' lightweight railcar, which in 1934 set a transcontinental speed record by covering the 1,015 miles between Denver and Chicago at an average at 124.9 km/h (77.6 m.p.h.).

Gresley's A4 Pacific was developed in response to the threat of German diesels that the management of the London and North East Railway were considering purchasing for their East Coast route from London to Newcastle. Germany were the European leaders in diesel traction; in 1933 the 'Fliegende Hamburger' ('Flying Hamburger'), a lightweight diesel train built by Wagen and Maschinenbau AG, had inaugurated the world's first 160 km/h (100 m.p.h.) passenger-carrying service. It was a version of this train that the LNER were contemplating buying.

Gresley's design pushed stream traction to its limits in order to outclass the relatively underdeveloped diesels. In so doing, the position of steam traction in Britain was so reinforced that its role was not seriously challenged until the 1950s. By this time, as Freeman Allen (1978, p. 47) notes, 'most of the world's major railways were dieseling or electrifying as fast as resources would allow. British Rail was the conspicious laggard.'

The British railway engineer's faith in steam, which moulded fast train innovation in Britain in the 1930s and relegated diesel and electric traction to the experimental realm for a further twenty years, is something quite separate from the evolutionary nature of rail developments *per se*. Electric and, particularly, diesel traction were quite compatable with such a situation. Indeed, when Britain did replace steam, from 1955 to 1968, the transition period was the shortest of any major rail network.

What is of particular note is that it was not rail management or the commercial and operational side of the industry that really decided the sort of trains and services to be run, but senior railway engineers. Their design ideology, and power to impose that ideology, was and remained strong. Steam was not only a technology in which British designers were clearly world leaders, but it also represented a greater control over design by senior railway mechanical engineers. They had less control over the development of diesel and electric engines, with much work resting in the hands of supplying companies. Diesel and electric trains were therefore a direct threat to the whole professional status and experience of these senior mechanical engineers.

For the purpose of this study it is not necessary to enter into a detailed examination of the political and cultural circumstances that led to a steam rather than a diesel train being developed to ward off the threat of German imports. But the key role that senior mechanical engineers played in determining the criteria for rail innovation is an aspect that has proved to be of continuing relevance through to today. It is not an aspect that is unique to Britain, for the relative roles and power of engineers, operators and management has been crucial to the way in which many nations have approached rail innovation.

High-speed steam: an irrelevant distraction?

Although technologically spectacular, the overall impact of the fast steam trains of the 1930s was slight. They could travel at 160 km/h, but shared tracks with trains travelling at little more than half this speed. As mentioned above, this played havoc with timetables, track utilization, and a variety of other operational criteria geared to a totally different order of train speeds. Also, the ride of these trains was decidedly uncomfortable. The suspension systems of the carriages (particularly the articulated bogies) could not cope with the high speeds, resulting in the carriages swinging and suffering from a high level of vibration. This problem necessitated extensive track and vehicle maintenance in order to achieve 160 km/h operations. By 1939 the railways were moving away from such prestige 'flyers'. They were good for publicity, but really did not make much economic sense.

By the late 1930s, the railways were in a bad way. Overall income had fallen from £49.3m in 1929 to £29.8m in 1938 (see Hamilton & Potter, 1985, pp. 22–5). Competition from roads for both passengers and freight was intense and the depression of the 1930s had particularly hit major rail customers such as heavy industry. Coupled with this, the railways operated under 'common carrier' regulations which did not apply to road transport. Under these, the railways were obliged to carry any goods offered by customers. This required staff and special stock to cope with disparate freight which generated an income well below costs. Given this context, the development of fast, up-market, trains did not represent a priority. It could well be questioned why they were developed at all when the main threat to passenger traffic was from cheap coaches and buses. In practice, the fast train developments of the 1930s constituted little more than a distraction from the real problems faced by the railways at that time.

Modernization and the retrenchment of Beeching

The Second World War pushed Britain's railways to their operational limits, and the damage and maintenance problems of this period, coupled with the low investment of the depression years, paved the way to nationalization in 1948 by the post-war Labour Government. Fast trains were by no means on rail's agenda at this time, however much engineers might have liked to build them. The A4 Pacifics were used for ordinary express duties, not as prestige 'flyers'.

None the less, investment continued to be in steam, reflecting the continued influence of senior mechanical engineers on train design. Whereas the role and scope of engineering altered little upon nationalization, that of management and operations was radically changed. The dominance of engineering within British Railways was further reinforced by successive upheavals in railway management in the late 1940s and early 1950s. At nationalization, the British Transport Commission (BTC) was established to co-ordinate and plan all the nationalized transport industries. The railways were only just beginning to settle down from this when the Conservative Party was returned to office. In 1951 the BTC was split up, with some sections (though not rail) being sold off to the private sector. Management was again reorganized and the BTC's planning and co-ordination objectives replaced by more competitive commercial criteria. Railway management, facing changing objectives and hardly settling down from one reorganization when the next came, was left in a weak position. The senior engineers had no difficulty in maintaining their ideas on what sort of engines they wanted to build and what sort of operations these trains should be used for.

By the mid-1950s, Britain was really lagging behind in the replacement of steam traction and rail modernization in general. Somewhat suddenly, the Conservative Government realized that something needed doing quickly for the railways to function at all effectively in modern Britain. Pressure from industry, which could see no real improvement in the service the railways offered them without a major injection of state aid, was probably the main factor behind the Ministry of Transport inviting the British Railways Board to spell out its ideas for rail modernization.

Up until the 1955 Modernization Plan, what investment there had been had been predominantly in steam (indeed steam trains were built until 1960). With signs of real government support, rail management was now in a stronger position and was headed by a new and very capable Chairman, Sir Brian Robertson. His *Modernization Plan* envisaged the rapid replacement of steam by diesel traction and the electrification of the East and West Coast main lines. Save for the trans-Pennine Sheffield to Manchester route, the only electrified lines at this time were for suburban commuting services.

The system of electrification proposed in the Modernization Plan was the use of 25,000-volt AC overhead lines, pioneered in France. At the time this was a new, innovative system and it was a bold decision to go for it, especially given the existence of an extensive 750-volt DC third rail network on the Southern Region of British Railways. In practice, the high voltage 25Kv overhead line system turned out to be one of the best features of the Modernization Plan, being more efficient and economical than other electrification methods. It is now the standard system for European railways.

It was recognized in the Modernization Plan (and the subsequent 1956 White Paper *Proposals for the Railways*) that electrification would be a long-term process, with projects taking fifteen years to be completed. Diesel was therefore to be used as steam's initial replacement. There was a brief dabbling with gas turbine power in the late 1950s, with three locomotives being adapted for passenger service. One was even a converted steam train! However, this option was never seriously pursued in Britain until the advent of the Advanced Passenger Train project in 1972.

Besides dealing with motive power, the modernization Plan identified other priorities for the creation of a modern railway system. The replacement of wooden sleepers and jointed rails by concrete sleepers and continuously welded track and the extension of coloured light signalling to replace the old semaphore system all featured. These were all very much in the evolutionary tradition of the railways, representing technologically gradual, though economically important, improvements. The most radical feature of the Modernization Plan was for freight, perhaps the railways' biggest headache. This involved the concentration of freight handling in a network of large modern depots, with the closure of one hundred and fifty smaller freight yards.

The railways were not alone among transport modernization projects in the mid-1950s. The concern that the railways were not capable of serving a modern economy and hence required an injection of state support was mirrored in other transport sectors as well. The mid-1950s saw the beginnings of the motorways programme (the M1 opening in 1959) and the further development of an internal air network.

Like many government projects of the 1950s, the Rail Modernization Plan was scaled down and subject to cuts, but it suffered more than other transport programmes. The development of the motorways and internal air services went ahead unabated. This was indicative of a state attitude that was to impose additional constraints upon rail investment in Britain.

Evaluating transport investments

Railways in most countries, even those under private ownership, are subject to state control and political pressure. This is because they are viewed as a transport service to the nation, not just an industry that can be judged purely in terms of an internal profit or loss. Such wider criteria vary from strategic military use and the aiding of industry to stimulating regional economic development and energy policies. For example, the development of the 'super-metro' 'Reseau Express Regional' (RER) around Paris in the 1960s was specifically linked to the development of suburban new towns and the decentralization of people and industry from the city. As a project on its own it was not commercially viable, but it played a key part in transforming the urban/economic system of the Paris region. It was as part of this that investment in the RER was evaluated.

Early arguments for rail nationalization emphasized such wider functions. For example, Winston Churchill supported the abortive 1918 attempt at nationalization because 'railways in private hands must be used for immediate direct profit, but it might pay the state to run the railways at a loss to develop industries and agriculture' (quoted in Hamilton & Potter, 1985, pp. 33–4). The rationale behind the 1955 Modernization Plan reflected a similar viewpoint, in that it was the role of railways in the economy as a whole that was the criterion for state investment, not the economic performance of the railways alone.

A number of techniques have been developed to apply such a 'systems approach' to transport planning, the most common of which in Britain is cost-benefit analysis, which is particularly used to assess road building. This involves the assessment of *all* benefits and costs involved in an investment regardless of who incurrs or receives them. So, for a road scheme, for instance, costs will not only include those incurred in building the road, but factors such as noise, visual intrusion, loss of agricultural production, etc. Benefits include time savings, reduction in congestion and accidents in bypassed towns, reduction in transport costs to industry, etc. Overall, a scheme is judged to be worthwhile if benefits exceed costs by a certain ratio.

A major source of contention with cost-benefit analysis concerns evaluating cash benefits against non-quantifiable factors such as noise, visual intrusion and other social as opposed to economic factors. This is because the technique relies on reducing all factors to a cash value. Other systems approaches concentrate on broad social and economic objectives and then consider what methods are needed to achieve them, which is the opposite approach to the way in which cost-benefit analysis examines the effect of a particular proposal upon an existing situation. Generally, transport planning in Britain today is very *ad hoc* and is little influenced by integrated systems. approaches.

Governmental attitudes as to the role of railways and the means of evaluating their use varies greatly in different countries. In Britain, the late 1950s marked a significant watershed for, almost as rapidly as the Conservative Government adopted a systems-like approach to rail investment, it shifted back to treating the railways in economic isolation, to be evaluated purely in terms of internal profit and loss. Wider criteria continued to have a stronger influence in some areas, notably suburban and some rural services (and as such these came to have regular subsidies), but overall the predominant attitude came to be that of viewing railway economics in terms of annual accounts and balance sheets.

This state attitude in Britain reflects the power of the other transport industries, particularly the so-called 'road lobby' of road building and motor transport industries together with their associated unions. Road interests had formed lobby groups from the turn of the century. The best-known of these are the Society of Motor Manufacturers and Traders (the SMMT), formed by the car manufacturing companies and distributors, and the Road Haulage Association, formed by road freight operators. These and other road transport interests combined to establish the overtly lobbyest British Road Federation (BRF). By the mid-1950s, in the wake of rising private car ownership and a rapidly growing road freight sector, the political influence of these lobby groups mushroomed.

The mid-1950s also marked a shift in power within the trade union movement. As traditional industries declined and manufacturing, including the car industry, expanded, so the power balance in the unions shifted towards the Transport and General Workers Union (T &GWU). This was the union to which the car workers and lorry drivers belonged and it generally supported the anti-rail stance of the road lobby. Added to these lobbying factors were two other aspects which encouraged the government to cut rail investment in the 1950s. One was that the car industry was seen as an important area of job growth to help compensate for job losses in traditional industries, such as coal, which alone lost 300,000 jobs in the 1950s and early 1960s. In hindsight, it is difficult to envisage how rail modernization could have had a serious impact on the growth of the British car industry. However, a political atmosphere in which the motor industry was seen as the main hope for generating jobs amid a post-war population boom shifted the focus for government action away from rail modernization.

The other factor was that the Rail Modernization Plan had been put together very quickly and there were some genuine reservations about the plan itself. The programme was very big and costs had risen from £1,200m to £1,500m. But what really tipped the balance against the railways was the appointment of Ernest Marples as Minister of Transport in 1959.

Marples at that time was the senior partner in a road building company, Marples Ridgeway, which had just finished building the Hammersmith Flyover, the first elevated road in London. As noted in Hamilton & Potter (1985, p. 52):

> Being the Minister of Transport, Marples was not permitted to retain the ownership of his road construction company. Such a personal interest in the outcome of this policy decisions could not be allowed. He passed his shareholdings on to his wife, which apparently he viewed as eliminating any personal interest or gain in his being responsible for Britain's roadbuilding programme!

Marples was determined to see through a popularist road building programme and at the same time set into motion an inquiry into the railways that was to lead to the Beeching Report and the subsequent axing of a third of Britain's railways.

The whole political atmosphere in the 1950s and into the 1960s was against the railways. Government policy was to reduce the role of the railways and their level of state aid. The criterion of the Beeching Report (1962) was that the railways should break even financially by 1970. This highlights the contrast between the investment criteria for road and for rail. The road lobby succeeded in getting state road investment evaluated using systems-type cost-benefit analysis, while discouraging such an approach to rail investment. This fragmented approach to British transport policy meant that, although rail modernization did proceed, and steam was totally replaced by 1968, there were severe financial constraints upon rail investment at a time when state aid to rail's competitors seemed almost unlimited. This attitude was a feature of both Conservative and Labour administrations.

With cost-benefit analysis used to assess road investments and commercial criteria for rail projects, far more road projects qualified for government finance than rail and hence rail's competitive position declined. An example of this can be seen in the series of toll bridges built in the 1960s/70s across the major estuaries in Britain—the Severn, Forth and Humber—together with the Dartford Tunnel beneath the Thames. The tolls charged do not even cover the interest charges on the capital borrowed to build them. If these key links in our motorway networks had been evaluated in the same way as railway investments they would never have been built.

Because of the methods used to evaluate the role of railways in our economy and society, Britain's railways receive far less investment funds than elsewhere in Europe. In a comparative case study (TEST, 1984), British Rail's 1982 investment of 11,000 European Currency Units (ECUs) per route-km contrasts with 42,000 for West Germany, 71,000 for the Netherlands, 103,000 for Switzerland, 15,000 for Sweden and 19,000 for France. Furthermore, whereas annual rail investment levels in these countries have either been maintained or increased over the period 1976–1982, that for British Rail dropped from a real total of 684m ECUs to 202m ECUs. Consequently, one reason why a number of improvements to the railway network successfully applied elsewhere in the world (e.g. new track, widespread electrification, etc.) are non-starters in Britain is because they cannot meet the internal profitability criteria that is demanded of them in this country. This will be considered in more detail later, when such projects are examined.

Constraints and stimuli

This brief and selective run through the history of fast train developments reveals a number of important constraints and controls over the innovative process. Firstly, the sheer scale and nature of a railway system as a whole means that it is technically and economically impossible to introduce an innovation that does more than develop one part of it at a time. As will be considered in the next section, the exact nature of the legacy of infrastructure can also be significant.

But on top of this general 'evolutionary' approach to innovation, a number of other factors can be identified. The relative power of 'actors' (or departments) within the railway industry is clearly crucial. The power of senior mechanical engineers in Britain and the way this was used to quash the serious development of diesel and electric traction until 1955 is of particular note. The contrast with Germany, which rapidly abandoned the further development of steam in the mid-1930s, could not be stronger.

The fact that both management and engineers saw speed as *the* area in which rail should compete for passenger traffic also deserves closer scrutiny. The evidence from both the United States and Britain suggests that, although improvements in speed were clearly an important way for rail to compete, the decisions to develop and run fast trains were based on a somewhat simplistic analysis of rail's markets.

But, finally, the role of government and the political power of the railways relative to the automotive industry has clearly come to be crucial in determining the resources that are available for rail investments. By the early 1960s, this had come to place a tight constraint on rail innovation in Britain.

The development of Inter-City Services

Despite government cutbacks, by the early 1960s the modernization of British Rail was well under way with the replacement of steam traction and old coaching stock by modern equipment. The main emphasis was on changing from steam to diesel and electric traction. Innovations were entirely 'defensive', simply seeking to fulfil rail's traditional roles in a cheaper way. Cutting costs was particularly emphasized in the 1962 Beeching Report, with modernization investments coupled with the closure of branch lines and the withdrawal of other railway services. Innovation was not orientated towards seeking new markets, but to consolidate old. The aim was simply to achieve savings in running costs and manpower and so make the railways more competitive.

Given such an attitude, speed was not seen as particularly important and train timings showed only a marginal improvement (if any at all) over pre-war steam operations. A number of express services averaged 100 km/h (62 m.p.h.), but overall averages were of the order of 80 km/h (50 m.p.h.). But some railway managers were advocating a more aggressive approach, especially given the growing impact of the new motorways, rising car ownership and expanding internal air services. Journey times by rail had improved marginally, but were nothing compared to the improvements achieved by the vast state investments under-way in road, domestic airports and aircraft developments. In America, as will be discussed in the next chapter, the combined effect of the airlines and rising car ownership had decimated the railroad's passenger business. There were larger social and economic forces coming into play than could be addressed by Beeching's cost-cutting approach. The transport market of the 1960s and beyond was going to change in a big way and the railways were going to have to change their product and approach.

Speed thus returned to the agenda of Britain's railways for the first time since the 1930s. It was on the same East Coast London–Newcastle–Edinburgh route that had been operated by the 160 km/h (100 m.p.h.) A4 Pacifics in the 1930s that 160 km/h operations returned to Britain in 1962. The A4 Pacifics were still operating as front-line equipment on this route and, under the 1955 Modernization Plan, were due to be replaced by electric locomotives. But this electrification plan had been axed as part of the 1959 Marples cutback in rail investment, so a lower-cost diesel modernization scheme was the only option open. However, the Eastern Region management of British Railways took the initiative and sought a high performance diesel locomotive capable of regular operations at 160 km/h (100 m.p.h.) to replace steam. The locomotive chosen was English Electric's twin-engined 3,300 hp (2,462 kW) *Deltic*, a privately developed design in which, until then, British Railways had shown little interest.

The object of using trains with a 160 km/h capability was to achieve an average journey speed of 120 km/h (75 m.p.h.), which it was considered would keep rail competitive with the motorways. In conjunction with the new locomotives, selective improvements to the existing track were planned so as to eliminate speed restrictions at junctions and through intermediate stations. A number of curves were also realigned in order to raise speeds. The idea of a 'threshold speed' at which trains could effectively compete with the car and air was not derived from any theoretical modelling concept (as will be discussed later in this chapter and in Chapter 6), but was simply based upon day-to-day operating experience. That speed was seen as the area in which innovations should be concentrated was a response to the fact that it was mainly in speed that the competitors to rail were making rapid improvements. This reaction was not unique to rail management in Britain. Other railways in Europe and elsewhere saw speed as the key to rail's future competitiveness. For example, an

Italian study in 1962 concluded that to remain competitive with air and the motorways, operational speeds in excess of 150 km/h (90 m.p.h.) were necessary (details of this are in Chapter 3).

The raising of average journey speeds on the East Coast route from 100 to 120 km/h was a relatively modest goal and did not involve any major technical advances, even though the *Deltic* was, at that time, the world's most powerful diesel locomotive. Achieving high speeds was a product of combining this train with track and signalling improvements, together with a timetable that effectively combined these fast trains with other traffic.

The accelerated East Coast service, using the new Deltic diesel locomotive (Figure 3), was inaugurated in 1962 and is generally seen as the turning-point for British Rail's post-war Inter-City services. This pattern of fast regular-interval trains showed that rail could compete with road and air and became the model for Inter-City rail operations throughout Britain to this day.

Figure 3. A Class 55 Deltic-hauled train on the East Coast main line
Source: British Rail

The one Inter-City route that was electrified in the 1960s was the intensively used West Coast main line from London to Birmingham, Liverpool and Manchester. This £163m scheme only just got through the 1959 cutbacks. In 1960 the Select Committee on Nationalized Industries rather unenthusiastically recommended that the scheme go ahead and the government was sufficiently doubtful to have work on the project stopped for a brief period in late 1960 (see Evans, 1969). Eventually work proceeded and in April 1966 160 km/h electric services from London to Liverpool and Manchester began to be followed a year later by services to Birmingham. These trains were hauled by a series (Class 81–86) of AC locomotives of 2,200 to 2,686 kW (Figure 4). Timetabling was similar to that introduced on the East Coast main line and modern signalling was also installed. Some minor track improvements took

Figure 4. A Class 86-hauled train on the West Coast main line
Source: British Rail

place, but these were of a smaller scale than those undertaken on the East Coast main line. Journey times from towns and cities along the electrified route and on feeder lines into it were reduced by between 20 and 30 per cent: for example, the Manchester to London time was reduced from 4 h 22 min to 3 h 15 min.

Again, none of the elements involved in the West Coast electrification could be seen as innovative in themselves. It was the application of a new mix of existing technology and operational methods that produced faster running speeds. Indeed, if anything was particularly innovative about the East and West Coast route schemes it was the timetabling used. All trains ran equally fast . . . unlike the prestige 'flyers' of the 1930s. Hence track utilization was good as there was no need to leave a half-hour gap (or more) in front of a fast train to stop it from catching up on a slower one, or the need to build extra tracks to allow for passing. This timetabling design also utilized the trains well, making the most of the limited investment in new stock.

The commercial impact of these two mainline schemes was considerable, given the relatively small investment and the highly evolutionary nature of the technical and operational innovations involved. Up until this time, Inter-City services had been suffering a decline and there were many in rail management who saw their job as that of overseeing the gradual demise of long-distance rail passenger services as railways retreated into a commuter and freight role.

By the late 1960s, however, passenger traffic on the West Coast main line had doubled and electrification was continuing to Glasgow (completed in 1975). It was clear that the railways could compete with road and air, even given limited investment funds, if innovations reinforced a market strength. For rail, speed was the obvious market strength to focus upon. Passenger surveys became more sophisticated and suggested that for Inter-City journeys of

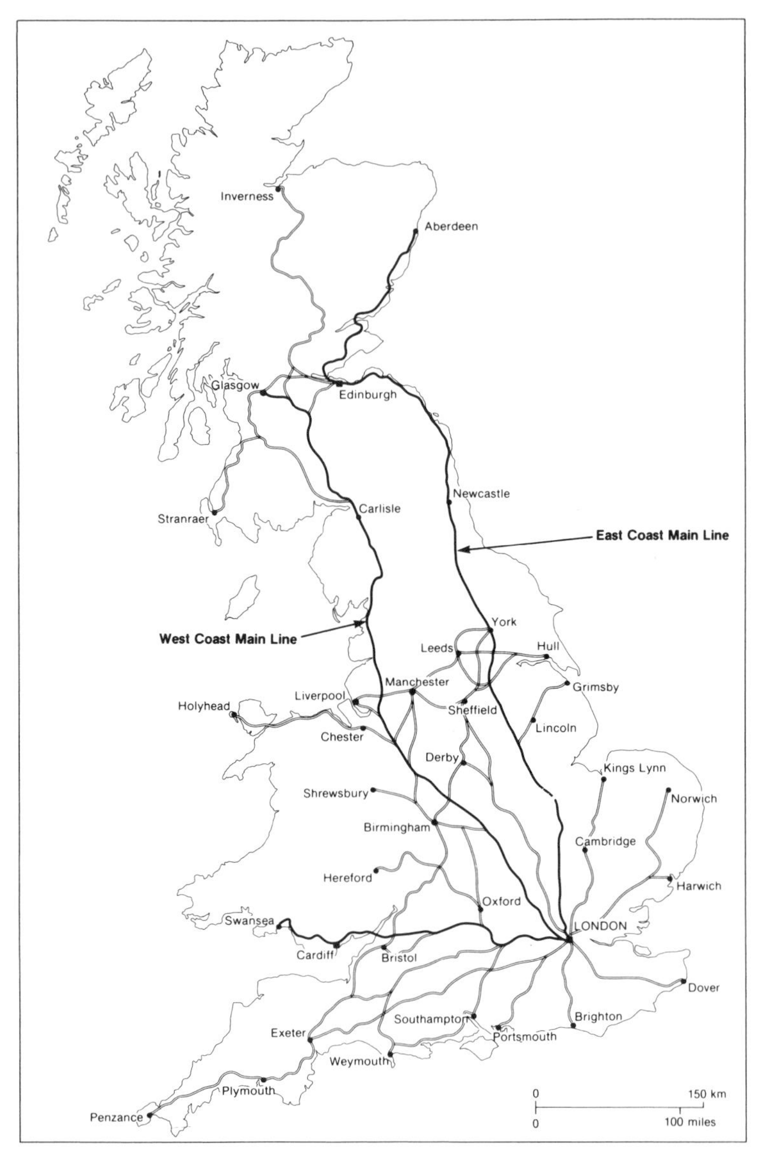

Figure 5. Principal Inter-City railway lines in Britain

Source: Potter & Roy, 1986, p. 11

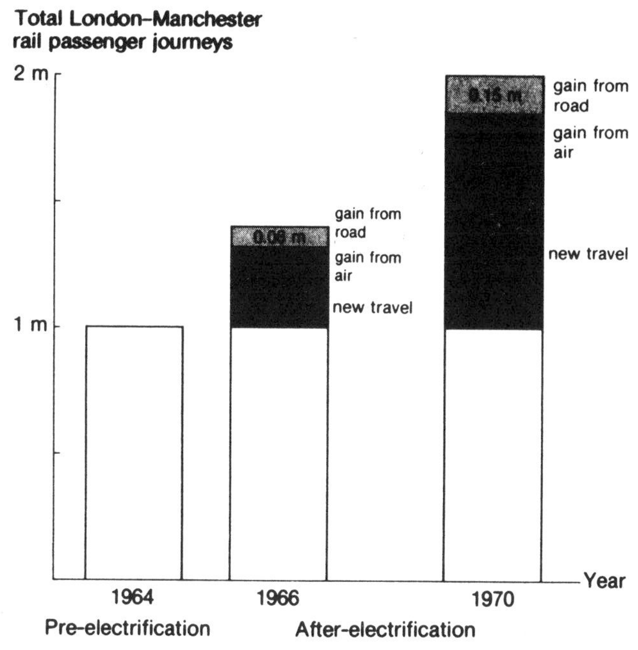

Figure 6. Growth of rail passenger traffic on the London to Manchester route, following electrification in 1966

Source: Potter & Roy, 1986, p. 12

up to three hours, rail could successfully compete with the car and air and even generate significant amounts of new travel. Cutting journey times to under three hours—a reasonable target on a relatively small island like Britain—thus became a major policy objective of British Rail.

As shown in Figure 6, for the West Coast Main Line electrification, the initial reaction to the improved service was a transfer from other travel forms to rail. In later years this shifted to growth from new traffic. Such a pattern would be expected when an innovation produces a 'one-off' advantage. Speed improvements on other routes yielded a similar pattern. An 8 per cent time reduction on the London–Plymouth run led to a 15 per cent traffic increase. A similar time reduction on London–Newcastle using the Deltic diesels produced a 20 per cent traffic increase. As this experience of fast train operations built up, an empirical model of the effect of higher speeds upon passenger revenue emerged. Basically, the experience of the Deltic and electrified West Coast Line operations suggested that for every 1 m.p.h. improvement in speed a one per cent increase in passenger revenue could be expected.

This relationship between speed and passenger revenue is commonly referred to as 'speed elasticity'. This is an adaptation of the economic and marketing concept of elasticity, which is a measure showing how much demand for a product will change in response to changes in the price or the quality of that product. The most commonly used measure is price elasticity, i.e. a measure of the extent to which demand will drop when a product's price rises, or demand will rise if the price goes down. Elasticity is the angle of the demand curve relating price and quantity (Figure 7). In its simplest terms, elasticity is the ratio between the percentage change in quantity demanded and the percentage change in price. More sophisticated methods of calculating elasticity exist which take into account that the angle of the curve will vary at different points on it, which take the form of a formula describing the whole curve.

For the railways, speed is a measure of the *quality* of a good (i.e. rail services). So speed elasticity is the ratio between demand and average journey speeds. To take the example given of the London to Manchester route, see Table 1. This example assumes a simple relationship between two variables. In his assessment of the London Midland electrification, Evans (1969) discussed the problem of isolating the effect of speed from that of other characteristics of a rail service. Speed was not the only improvement made to the quality of British Railways Inter-City services in the 1960s. This period also saw the introduction of considerably improved coaches, many

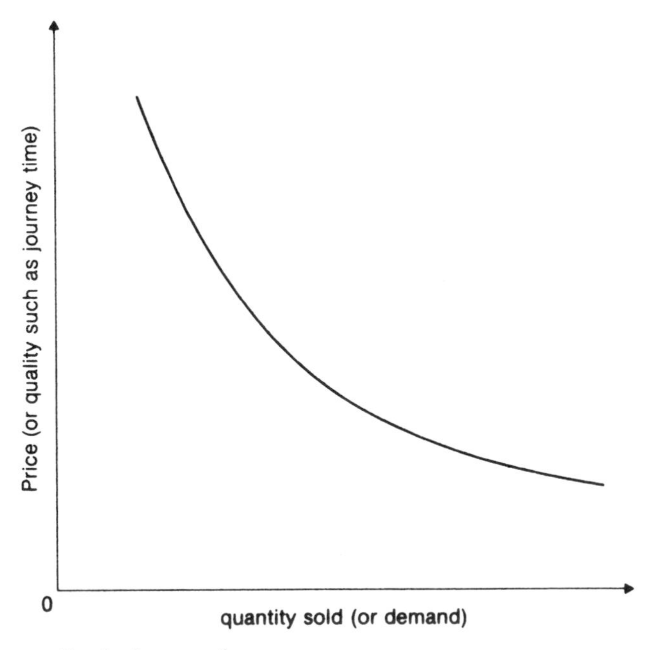

Figure 7. A demand curve
Source: Potter & Roy, 1986 p. 12

with air conditioning. By 1976, 45 per cent of Inter-City trains had air conditioning, which rose to 90 per cent by 1984. A number of major stations were rebuilt and a whole new marketing emphasis was given to long-distance passenger operations. The term 'Inter-City', to describe these services, was introduced in 1966 as part of this marketing revamp.

Table 1 The London to Manchester route–speed elasticity and demand

	Old amount	New amount	% change	Elasticity
Average speed	70 km/h	97 km/h	+39	$\frac{+39}{+40} = +0.98$
Passengers	1.0 m	1.4 m	+40	

Although the role of these other factors in generating passenger traffic was acknowledged, it was widely believed that the most crucial factor was speed, and hence speed elasticity studies came to play a central role in Inter-City railway planning in Britain and many other countries. Today it is considered that the railway planners of the 1960s overestimated the role of speed in generating new business. None the less, it was the introduction of fast trains that sparked off a new approach to British Railways' Inter-City services and got the railways to turn aside from the negative attitudes of the Beeching era.

3 The Shinkansen

Britain was by no means alone in its interest in fast train developments in the 1960s. In many industrialized countries motorway building, rising car ownership and competition from the airlines resulted in a similar reaction. Speed came to be viewed as the main way for rail to hold its own; but how to achieve high speeds was another matter. Up until the early 1960s, the method used in Britain was the norm: improvements to existing track, electrification and the gradual development of faster locomotives and rolling stock. But with the opening, in 1964, of Japan's first Shinkansen Line (the Tokaido Line, between Tokyo and Osaka), came the realization that more radical alternatives existed—alternatives that did not just involve marginal improvements in speed, but raised operational speeds by an order of magnitude.

Japan's 210 km/h Tokaido Shinkansen was the pioneer of modern fast railways and the circumstances that led to it being built were virtually unique. The decision to build a totally new passenger line owed little to air competition or the threat of the car. It was simply that the existing line was heavily congested and additional track capacity was badly needed.

The coastal strip between Osaka and Tokyo forms a tight, 300 km-long 'corridor', along which travel flows are highly concentrated. Today, the Pacific coast corridor contains over thirty-seven million people. Japan's post-war economic boom put intense pressure on all forms of transport within this Pacific coast corridor, so by the mid-1950s new roads, motorways and airports were all being developed, as well as the new Tokaido railway line.

Nowhere else in the world is there such as intense concentration of people along a particular route. Japan's Pacific coast corridor averages 123,000 people per kilometre. Parts of the north-east United States and Germany have route corridor population densities of around 30,000 to 70,000 per route km and a number of other industrialized countries (including Britain) have corridors with population densities of 10,000 to 20,000 per route km, but none approach the unique geographical distribution of Japan's population.

Plans for a new Tokyo–Osaka railway had been prepared in the late 1930s and even then it was earmarked for high-speed operations. Using both dc electric and steam traction, running speeds of up to 160 km/h were envisaged, with an average speed of 130 km/h between Tokyo and Osaka. The onset of the Second World War halted the project, but some of the land already acquired was incorporated into the post-war Tokaido line plans.

Given the need for new capacity, the key decision in planning the new Tokaido Shinkansen (Shinkansen simply means 'New Trunk Line') was to concentrate all express services on the new route, leaving the existing line to cater for freight as well as passenger services to smaller towns. Perhaps the most unusual decision was to make the new line *entirely* separate from the rest of the railway network. The existing Tokyo–Osaka Tokaido line, like most of Japan's railways, was built to a narrow 1.067 metre gauge. The new line was entirely isolated from all other railways by being built to the 1.435-metre standard gauge. Hence it has its own trains that cannot run on any other tracks in Japan. The reason for this was technical, in that the design speed of 210 km/h required a wider gauge for safety. Given the narrow population corridor in which demand was concentrated, the restriction of the trains to the single main line was considered of little consequence.

Despite the radical concept of a dedicated passenger railway operating at speeds a third higher than elsewhere in the world, the technology of the Shinkansen was not innovative. Wickens (1983, p. 95) describes it as 'extremely well-engineered conventional steel wheel on steel rail practice.' The new track is electrified, using the standard 25 kv overhead line method and includes substantial elevated sections, mainly because, in densely populated Japan, this is cheaper than building a host of individual road bridges.

Right from the beginning, the Tokaido Shinkansen was intended to be an ultra-express service with the trains running as fast as existing technology would allow. In 1957, a Japanese Technical Research Institute paper held out the prospect of a three-hour Tokyo to Osaka journey time, involving a top speed of 210 km/h. This caught the public imagination and played a major role in persuading the government of the day to fund the project through a World Bank loan.

Although a third faster than any train then in operation, 210 km/h, represented about the fastest that was technically achievable in regular passenger service at this time. As will be considered in detail in the next section, there were two major factors which imposed a limit on rail operational speeds. The first was the unstable vibration of a rail vehicle's wheels at high speeds (called 'hunting'). This could be minimized by the use of specially designed wheels which, however, did not curve well and hence required a new track with very gentle curves.

The second factor was track damage. The French had undertaken a series of high-speed runs in 1955, setting a world rail speed record of 331 km/h in the process. But the effect upon the track and overhead power lines had been disastrous. The rails were twisted and unusable and the overhead wires and the train's pantograph had both been damaged by the pantograph bouncing on the overhead wire, causing the electric current to arc over the resultant gap. The sheer force on the track and overhead wires imposed a speed limit on the Shinkansen trains. Even at

210 km/h, wear on the track and overhead lines is exceptionally high and the Shinkansen is closed every night for maintenance.

Construction of the Tokyo–Osaka Tokaido line started in April 1959 and passenger services began in the summer of 1964, reducing the travel time between these cities from 6 h 30 min to 3 h 10 min. Traffic increased by 66 per cent in the second year of operation and an extension of the line (called the Sanyo Shinkansen) was authorized. This opened in stages, from 1972 to 1975, doubling the length of the original line. This route now carries 124 million passengers a year. A measure of the concentration of traffic on this corridor is that the world's next busiest route, the TGV line from Paris to Lyon, carries only thirteen million passengers per year.

The next stage in the development of the Shinkansen marked a significant break from the first. The rationale for the Tokaido line was basically that extra track capacity was needed and, given this, a logical way to provide it was as a

Figure 8. A Shinkansen train at Kyoto Station

dedicated fast passenger line. But the development of this line, and (more particularly) its extension, the Sanyo line, displayed the social and economic impact of fast passenger railways upon the less developed regions of Japan. As Takayama notes, 'Perhaps the most important (impacts) were the social and economic effects of the new line. Opportunities for industrial and housing developments opened up along the Tokyo–Hakata corridor and its hinterland; land once of little use was rapidly exploited.'

Japan's economy was booming, yet the sheer concentration of population and industry in the tight Pacific coast corridor was causing major problems of congestion. The potential of the Shinkansen to stimulate regional development led therefore, in 1970, to the National High Speed Railway Act, incorporating the concept of a national network of high-speed lines. Two routes were selected for immediate development: the Tohoku line linking Tokyo to Morioka and the Joetsu line from Tokyo to Niigata. The primary aim of both these lines was to underpin the economic development of the areas through which they passed. From being a necessary consequence of the congested urban concentration of Japan, the Shinkansen became a way of tackling this problem.

Construction work began on both these lines in 1971. They were due to be completed in 1976 but strong (and often militant) opposition from local residents and interest groups delayed progress. The main concern was noise and vibration from the elevated track, particularly as the lines (because of their unique gauge) had to be built through existing areas with houses adjacent (and often under!) the track. A directive issued by the Japanese Environmental Agency in July 1975 specified tougher noise level standards and as such a range of anti-noise features was incorporated into the track design. At the Tokyo end of the Tohoku route, Japanese National Railways faced major problems of land acquisition. In the end they had to build a temporary suburban terminus at Omiya in order to open

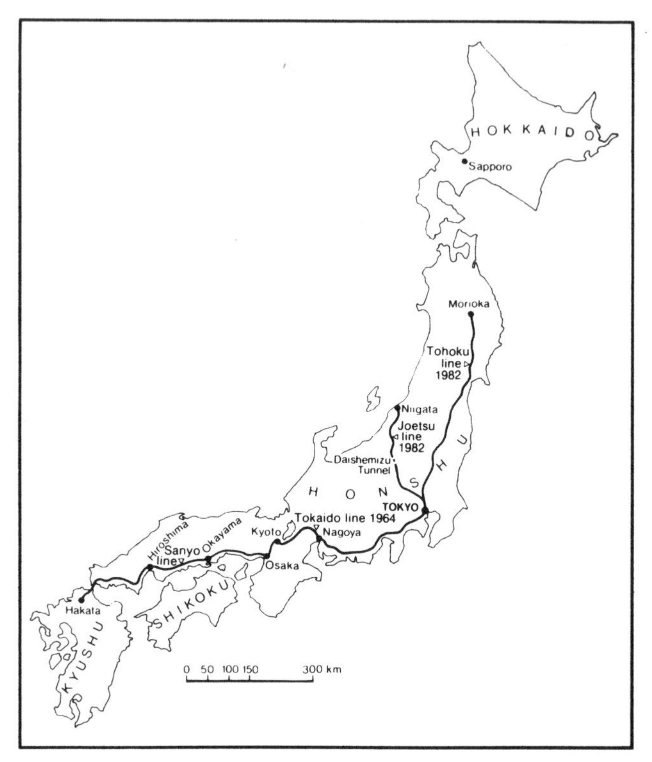

Figure 9. Japanese National Railway's Shinkansen Network
Source: Potter & Roy, 1986, p. 14

the line to traffic. Both the Tohoku and Joetsu lines eventually opened in 1982 and now carry around twenty million and thirteen and a half million passengers per annum respectively. The link to Tokyo's city centre Ueno terminus finally opened in March 1985.

The Shinkansen influence

The immediate success of Japan's Shinkansen experience had a major impact on the thinking of railway management in many countries. In Britain it helped British Rail to shrug off the negative Beeching era, in France and

Germany the concept of building new high-speed lines began to be seriously canvassed, but it was the Italians, with plans for a new Rome to Florence *Directissima* who led the way.

The Italians had been the first country in Europe to set about building new fast railway lines in the inter-war period. As early as 1913 they embarked on a new line to shorten the distance between Bologna and Florence (which took twenty-one years to build, finally being completed under Mussolini in 1934). A second *directissima*, a new line shortening the Rome to Naples route, was opened in 1927. Once electrified in 1939, 160 km/h (100 m.p.h.) operations resulted in an average speed between Rome and Naples of 118 km/h (72.5 m.p.h.).

As mentioned in Chapter 2, like British Rail, the Italian Railway management saw speed as the main way of competing with road and air. In 1962 they undertook a study which concluded that they would have to aim at achieving speeds in excess of 150 km/h (90 m.p.h.) to ensure that rail would remain competitive with air up to 1,000 km (625 miles). This involved ordering new, fast, lightweight electric trains which could only be fully exploited on the new lines. Speeds on the Rome–Naples *Directissima* were raised to 180 km/h (112 m.p.h.) and Italian Railways began to canvass for a new *Directissima* between Rome and Florence. In the wake of the success of the Shinkansen, the Rome–Florence *Directissima* was authorized in 1969.

In the United States, the long-distance rail passenger market had all but collapsed. The airlines had mopped up the medium to long-haul traffic while the vast growth in car ownership coupled with equally vast state expenditure on expressways had decimated short distance services. It is little wonder that many railway operators looked at the United States in prophetic horror, fearing a repetition once car ownership and airline development had reached a similar stage in their own country. The stark contrast between the American and Japanese railway experience sought to reinforce the influence of the Shinkansen in the thinking of railway management throughout the developed world.

In the United States itself, the fortunes of the railways turned with the passing of the High Speed Ground Transportation Act in 1965. Like the development of the later Shinkansen lines, the purpose of this Act was to contribute towards the relief of urban congestion, in this case the 'megalopolis' of cities along the north-east United States, from Baltimore through to Philadelphia and New York, down to Washington and Boston. A joint committee was established, involving the Pennsylvania Railroad, Federal officers and consulting engineers with the aim of upgrading the New York to Washington route to near-Shinkansen speeds. Upgrading the existing line for up to 200 km/h (125 m.p.h.) running was the target and it was planned to have the new 'Metroliner' stock in service by 1966.

This time-scale was hopelessly ambitious and the money for upgrading the track was insufficient to rebuild it for sustained 200 km/h operations. The fleet of Metroliners was hurriedly built and suffered from a host of technical faults, mainly in the traction control and braking systems, although there were also problems with the stability of the current-collecting pantograph at high speeds. Eventually they began to creep into service, from 1969, although it was another two years before a full 200 km/h service was in operation between Washington and New York.

The Federal government were really trying to obtain the results of the Shinkansen without incurring anything like the costs. The strategy of upgrading an existing line was certainly sound. The New York to Washington route had spare capacity and the track alignment was relatively good for speeds up to 200 km/h. The main problem was in the combination of a complex organizational structure, the desire for almost immediate results and a gross underestimation of the technical development required. Freeman Allen (1978) commented:

> Far too little of the resultant Metroliners' components had been put through the essential preliminary mill of practical evaluation in the tough railroad environment. Above all, it was soon obvious that the Metroliners themselves had been pointlessly over-designed; such infrastructure improvements as the budget would cover were insufficient to accommodate the Metroliners' maximum performance, even if that had been trouble-free.

In the end it had to be admitted that the 1966 plan had set a target that was unachievable with the resources allocated to it. The track was just not up to 200 km/h standards and, in 1977, work began on a $900m investment to rebuild the whole Boston–New York–Washington route in order to permit sustained 200 km/h running. In the end, some $2,500m was spent over ten years to upgrade this route.

The Shinkansen legacy

As the world's first dedicated fast passenger railway line, the Shinkansen was to have a lasting influence on railway management thinking. It presented a clear, perhaps simplistic vision of a long-term and expanding future for passenger railways. In the decade following the opening of the Tokaido line in 1964, managerial and political attitudes towards investment in fast passenger trains became more positive, with a number of plans for new lines and the upgrading of old routes to Shinkansen type speeds.

However, rather than simply imitating the Shinkansen, two countries sought to advance beyond it. In 1972, the prototypes of three fast passenger trains were unveiled: France's Train à Grande Vitesse (TGV) and British Rail's

Advanced Passenger Train (APT) and High Speed Train (HST). The approaches adopted involved strikingly contrasting methods. These methods, and the reasons for their adoption, form the basis of the next three chapters.

4 The concept of the Advanced Passenger Train

A 'scientific' approach to rail vehicle design

Up to the mid-1960s, the development of fast trains in Britain had taken place under the steady evolutionary pattern of the railway industry. Unlike in Japan, there was no one point where can it be said that a particular innovation had resulted in the fast and attractive rail services which, by 1967, were having such a market impact. The approach had been one of gradual improvement in a number of components making up long-distance passenger services: track and station improvements, faster locomotives, new carriages plus the better marketing of the product.

This evolutionary, empirical design approach is characteristic of most established medium and heavy engineering industries and is commonly referred to as the 'cut and try' approach. Typically, evolutionary design involves the following steps:

- DESIGN
- BUILD
- TEST
- MODIFY
- INTRODUCE INTO SERVICE

'Cut and try' is a slow but sure approach to evolutionary design improvements, but one that tends to preclude the possibility of radical innovation. It always takes existing designs and components as a starting-off point.

But in the mid-1960s, this traditional evolutionary design process was challenged by a radical research-based 'scientific' approach to rail innovation. This was the brainchild of an Advanced Projects Group in British Rail's Research and Development Division, who in November 1966 proposed the development of a train, the performance of which went well beyond anything contemplated by the operational side of British Rail. It took two years for the British Railways Board and the Ministry of Transport to take this concept seriously, but they eventually agreed to fund the development of an experimental train on a 50:50 basis. With this decision, in November 1968, was born the 250 km/h (155 m.p.h.) Advanced Passenger Train.

Rather than simply imitating the Shinkansen, the APT represented an entirely different approach to achieving, and exceeding a Shinkansen-type performance. The Shinkansen, and its immediate imitators, had all been designed and developed in the 'cut-and-try' tradition of railway design that had existed up to that time. The APT concept emerged from a programme of fundamental research into railway vehicle dynamics that required the solution of several theoretical problems which until then had been poorly understood by railway engineers. It was not the further development of an existing design, but sprang from basic research into the fundamental aspects of wheel/rail dynamics and rail bogie design. This required a 'scientific' approach involving much prior calculation, modelling and simulation, laboratory and full-scale experimentation before an attempt could be made to 'design—build—test—introduce into service'.

The capability to apply a research-based, scientific approach to rail innovation was the product of a substantial reorganization of the research facilities within British Rail which included the establishment of the Railway Technical Centre (RTC) at Derby in 1964. The British Railways Board wished to give a higher priority to research and to establish a central research authority for the railways. Until then, research had largely been viewed as a service to a particular department with a particular task in mind, such as testing paints for durability or improving the riding of rolling-stock. In line with the overall evolutionary approach, research in British Rail had been essentially 'defensive' in nature, aimed at the improvement of existing rail equipment. Freeman (1982) identified six basic strategies to innovation, of which the most common in industry is 'defensive' or 'immitative'. This he defines as 'concerned mainly with minor "improvements", modifications of existing products and processes, technical services and other work with a short time horizon.' (Freeman, 1982, Chapter 8). The establishment of the Railway Technical Centre (RTC) resulted in a significant shift towards a more 'offensive' strategy, with the research and development sections of British Rail seeking to establish themselves as the initiator of innovation. The purpose of the Research and Development Division ('BR Research') was (and still is) to act as a strategic organization, undertaking basic research on project ideas to be implemented by the appropriate engineering department and the workshops (or an outside contractor) once their viability has been proved.

The old approach to research, as a service to evolutionary developments, by its nature totally precluded anything but an evolutionary approach to innovation. But the co-ordinated and scientific approach of the Railway Technical Centre laid the foundations for radical innovation in rail technology to occur. At the Railway Technical Centre were located the design and development functions of the Mechanical and Electrical Engineering Department as well as the Research and Development Division and the British Rail Engineering Ltd (BREL) workshops. Thus, as Johnson & Long (1981) summarize the position:

> The Centre houses staff concerned not only with research, but also mechanical engineering design and development, workshops and supplies. The creation of

the centre has greatly facilitated the transition of projects throughout the stages from research through design to development and manufacture. Also, it has brought scientists into daily contact with engineers and so benefited mutual understanding. This unique blend of theoretical and practical expertise made Derby a focal point for world-wide interest in railway technological development.

The kind of fundamental research carried out at the RTC has included examining track problems, the riding characteristics of freight and passenger vehicles, the effects of vibration, fatigue and stress limitations, rail–wheel interaction and the effects of higher speeds on track and bridges.

Among the first projects to be undertaken was an investigation into 'hunting'—the unstable vibrations of wheel-sets at high speeds. Railway wheels are cone-shaped, since, as was discovered in the early years of the railway industry, this enables them to steer round bends. However, this steering motion is sinusoidal: in other words, it overcompensates for bends and track irregularities and sets up an oscillating motion from side to side. Up to a certain speed, this lateral oscillation is minimal and the ride of the vehicle is stable. But beyond this 'critical' speed, hunting occurs, with the wheel-set rapidly oscillating between the flanges. If speed is further increased, the ride of the train becomes more and more unstable, leading eventually to derailment. The critical speed at which hunting occurs depends on the type of rail vehicle in use, the quality of the track and how worn the wheels of the train are. At the time this research was undertaken, hunting limited passenger train operations to about 160 km/h (100 m.p.h.), but for two-axle, short wheelbase freight wagons the critical speed was much lower and hence many freight trains ran at under 70 km/h (44 m.p.h.). With the raising of track speeds for passenger trains through the 1960s, these slow freight trains were causing major timetabling and operational problems. But attempts to run freight at faster speeds had resulted in a spate of derailments which were blamed on hunting.

Freight vehicle design was therefore a major constraint on fast passenger train operations. Unlike in Japan, British Rail had a surplus capacity of track. This meant that mixed freight and passenger operations were necessary and 'new build' options were economically and politically out of the question. This situation therefore focused research effort on to resolving the technically complex hunting problem. In Japan, since new track was needed anyway, the technical problem of hunting had been sidestepped. For improvements in speed to be achieved on existing track, as was the case in Britain, the hunting problem had to be resolved.

Although hunting had been observed and research into its causes undertaken since the 1840s, empirical 'cut and try' approaches had not made much progress. There are so many interrelated factors involved in wheel/rail dynamics and bogie design that the problem was just not amenable to an evolutionary research approach. Only one way had been discovered that raised critical speeds. This was to use wheels of a low cone angle in order to reduce the tendency to unstable oscillations. However, low cone angle wheels reduce the steering ability of a wheel-set, so this imposes further major constraints. Trains can only be used at high speeds on specially aligned track with no sharp bends. Even so, critical speed is only raised from around 160 km/h to about 210 km/h.

Trains with low cone angle wheels can run on ordinary track but their curving performance is less satisfactory than ordinary trains. In addition, the use of low conicity wheels requires a lot of careful maintenance. The wheels themselves require re-turning about every 32,000 km (20,000 miles) in order to maintain their profile, otherwise hunting recurrs as the wheels wear against the track, causing their cone angle to become greater. Broadly speaking, this was the method adopted by the Japanese on their 210 km/h Tokaido–Shinkansen line, opened in 1964.

Even before the opening of the Railway Technical Centre, BR Research had already developed a theoretical computer model of axle, wheel, rail and suspension behaviour in order to examine the complex dynamics that caused hunting (for details of this work, see Wickens, 1971 and 1977. A summary of this appears in Potter & Roy, 1986 pp. 18–19). BR had aquired a digital computer as early as 1957 and was the first railway in Europe to use computers for scientific and engineering applications. Indeed, Dr Sydney Jones, who became head of BR Research in 1964, considered that 'the scientific approach to design was only possible after the availability of powerful electronic computers and fast acting electronic measuring devices to observe the effects of change.' (Jones, 1973, p. 52). Both these elements were involved in the computer models and subsequent empirical testing of rail bogie suspension designs.

This work resulted in a new bogie suspension design, incorporating springs and bars controlling lateral as well as vertical movements. This was tested in an experimental four-wheeled freight wagon, the type of vehicle that had been most susceptible to derailment by hunting. This wagon ran successfully at speeds of up to 225 km/h (140 m.p.h.) on a test bed at the RTC and at 160 km/h (100 m.p.h.) on open tracks. A fully developed design was incorporated into ten freight vehicles which ran in Scotland to determine the real-life behaviour of this innovation. After 120,000 miles there were no hunting problems and the wheels did not need re-turning. Indeed, the wheel design adopted had the profile of a *worn* wheel. This was because, as conical wheels deform on the track, the conicity becomes greater and hunting oscillations increase. By adjusting the suspension design to cope with a worn wheel profile, the effect of wheel wear on hunting is minimized and the wheels themselves only need re-turning after a large amount of wear.

Hunting had been an almost total barrier to commercial rail travel above 160 km/h. The capability of trains running well above the speed had existed since the 1930s, but the quality of track and level of maintenance required to keep hunting vibrations down to a safe and comfortable level had meant that these trains could not enter regular passenger service. The work at the RTC on suspension designs in the early 1960s made it feasible to run fast trains with worn wheels on imperfect, curved track. Speeds of up to 320 km/h (200 m.p.h.) were referred to as being entirely realistic on conventional railway tracks. The technical barrier to commercial high-speed rail operations had been broken, or at least raised sufficiently for it to be no longer of relevance.

Technological Rivals of the Advanced Passenger Train

This raising of the speed that could be commercially attained on conventional railway tracks had major implications for other fast ground transport innovations that were under development in the 1960s. One which got as far as a working research prototype by the early 1970s was a tracked hovercraft, powered by a linear induction motor. But this government-funded project was cancelled in 1973 after a total of £3.5m development money had been spent on it.

The tracked hovercraft was entirely technically feasible but it suffered from a magnified version of the financial problems experienced by the railways in this country. There was not the political will to see large sums of money invested in inter-city public transport and there was a definite fear, following the vast overspend on Concorde and other government funded projects, that this could occur again on another technologically spectacular scheme. But what probably clinched the fate of the tracked hovercraft was the fact that it required investment in a whole new network of track and vehicles in order to obtain any return on the initial investment. With the realization that redesigned 'conventional' trains running on existing track could achieve city-centre to city-centre journey times not far short of the hovertrain raised the question as to whether such an investment in new track was justifiable. A government-sponsored study examined the economics of investing in the APT, the hovertrain and in vertical take-off airliners. The result was that the technologically least advanced of the three, the APT, produced the highest return on capital.

The Vice-Chairman of the British Railways Board later commented:

> Much has been talked about tracked hovercraft and magnetic levitation. These might be good ideas in theory, but when it comes to nuts and bolts, the cost of equipping the country with an entirely new system of guided transport would be colossal. In steel wheel on steel rail we have a system that has proved itself capable of considerable development, and which has by no means reached the end of its stretch. [Lawrence, 1977, p. 262.]

This raises a question of continuing importance. The least technologically advanced transport system was the most economically viable. The application of technology was constrained by economics and the most important economic factor was the existence of a comprehensive railway network. Any development that required the construction of a completely new network of routes would have to have vastly more traffic-generating potential than one which was able to use the existing infrastructure.

Although the railways are a clear example of this economic constraint on technological development, this phenomenon can be seen in other sectors of the economy.

The lack of progress in developing electric road vehicles is another case in point. Although the recent fall in the price of oil has led to fading interest in electric vehicles, there is also a great infrastructural constraint on their development. For example, a study undertaken by the Open University Energy Research Group showed that, once oil reserves decline, the most efficient way to use remaining coal supplies would not be to produce petrol (the technology for which is well established), but to develop electric cars instead. However, this would require a totally different manufacturing, servicing and support infrastructure (including, for example, battery changing stations on motorways). Although once such an investment existed it would be more efficient than maintaining the petrol-engined car, the shift in technology involves such a massive change in the whole of the transport system as to render such a change highly unlikely. Instead, it seems that further refinements to existing car technology will continue and, for the foreseeable future, the most radical development we will witness will be in diesel cars and the use of electric vans for urban delivery duties, neither of which have substantial 'systems' effects.

Where a large supporting infrastructure exists that is geared to a particular technology, be it the 'steel wheel on steel rail' train or the petrol engined car, any alternative which needs an infrastructure system of its own would have to represent a massive improvement in either cost or performance in order to challenge the status quo. Compared to what conventional railways could do once the hunting problem was solved, the tracked hovercraft was judged not to have that massive advantage.

Despite this, the technology of the tracked hovercraft was not abandoned. By the time the project was cancelled, interest was shifting towards the use of Magnetic Levitation (MAGLEV) rather than an air cushion for such linear induction powered trains. MAGLEV uses an electromagnetic field to cause the train to rise a small distance

Figure 10. The Birmingham MAGLEV
These small MAGLEV cars operate a shuttle service between Birmingham Airport and Birmingham International Station

above the metal track. British Rail attempted to take over the tracked hovercraft project as a research facility, but this was vetoed by the Department of Transport. However, they continued to develop the tracked hovercraft's single-sided linear induction motor and have maintained an active interest in MAGLEV for small-scale applications and as part of the suspension design of conventional rail vehicles. Together with GEC and Brush, British Rail established a consortium which built the first operational MAGLEV line in the world, opened in 1984, linking Birmingham International Station to the new Birmingham Airport terminal (Figure 10).

Potentially, MAGLEV vehicles can travel at very high speeds, but in Britain, with journey distances relatively short and a large existing rail infrastructure, it is difficult to envisage an economic case for its use. In other countries, particularly if there are routes with a high traffic potential without an existing rail link, or where additional capacity is needed, the case for MAGLEV is stronger. There is a further discussion of this in Chapters 9 and 10.

From research concept to the APT project

As mentioned above, it took two years for BR Research to persuade the British Railways Board to support and fund the APT project. Eventually, in 1968, with an agreement from the Ministry of Transport to pay 50 per cent of the development costs of an experimental train, funding was agreed and the APT project got off the ground. That BR Research and the BR Board member responsible for research, Dr Sydney Jones, had to spend two years obtaining the go-ahead for the APT is indicative of the sort of internal and organizational politics present in BR at that time. On the commercial side, British Rail was just emerging from the negative retrenchment attitudes of the Beeching era and so the concept of such a technologically spectacular train as the APT was difficult to comprehend. But added to this was the considerable antagonism of traditional engineers in the Chief Mechanical and Electrical Engineer's Department (CM &EE). They believed the APT to be totally impractical and did not want to see resources wasted on such a project, particularly in an area that they felt was really theirs! This antagonism towards the APT really reflected a deeper antagonism towards the purpose of BR's Research Department and its shift from having a 'defensive' to an 'offensive' role. Johnson & Long particularly mention this in their study *British Railways Engineering: 1948–1980*:

> In the mid-1960s a deep division existed regarding the purpose and use of research in British Railways engineering. J. F. Harrison, as Chief Engineer (Traction

> and Rolling Stock) was heavily involved with such problems as standardizing main-line diesel designs, absorbing the English Electric Deltics into the fleet and technological evolution in a number of fields. A. N. Butland, Chief Engineer (Ways and Works) . . . hardly felt the need for the help of the Research Department unless it confirmed his own views, expressed with such vigour, on the permanent way (i.e. track) of the future. Those who were present when Jones was in disputation with these two engineers well remember the occasions, as neither of them found it easy to suffer even wise men gladly when in pursuit of departmental objectives.
>
> In effect, the Chief Engineers thought that the efforts of the Research Department should either be associated with these objectives or engaged in pursuing the possibilities of the less conventional forms of guided transport which were attracting a great deal of public attention at this time. Jones, however, felt very strongly that the traditional system of steel wheel on steel rail had a great deal of untapped potential if it were only to be subjected to fundamental re-thinking. [Johnson & Long, 1981, pp. 453–4.]

Eventually, Sydney Jones got the support of the Board and before long the operational passenger business side of BR became very interested in the APT concept. The 'offensive' concept of a research department had won, but was not necessarily accepted by all concerned. Johnson & Long considered that this reflected a 'creative' side to internal debate and division:

> If these opposed views had been held by weaker personalities, the results could have been unfortunate in any industry in comparable circumstances. In British Railways the debate was creative in that the engineers involved were stimulated to achieve results either by the process of evolution or by returning to first principles. [Johnson & Long, 1981, p. 454.]

This is, perhaps, a rather optimistic interpretation of the outcome of the internal wranglings over the role of research in British Rail and the APT project in particular. Although the support of the Board was achieved and the APT project went ahead, the internal rivalries between BR Research (especially the APT project) and traditional railway engineers were modified and lessened rather than eliminated. Some engineers became very supportive of APT while the scepticism of others remained.

Internal rivalries are typical of any large and complex organization and are often highlighted in a case, such as the APT, when an innovation threatens current practices and ways of operating. The APT was the first big project based on the new 'offensive' and scientific approach to rail innovation. Although the project did proceed there was a very real awareness among the project team that every mistake would be carefully observed and, if they were not careful, exploited by the evolutionary 'cut-and-try' school. The threat of the project being axed was a very real one and it probably would have been axed without the presence of Jones and other APT 'product champions' on the British Railways Board. The fact that the internal rivalries were not far below the surface, particularly in the early years of APT, very much laid the foundations for some of the trickier management problems that the project was to face.

The design concept of the Advanced Passenger Train

Although the main technical barrier to fast train operations on conventional track had been eliminated, the APT had other constraints to work within which soon led to further technical challenges and innovation requirements. The concept included the acceptance that major expenditure on railway track and other infrastructure, as required for Shinkansen, the Italian *Direttissima* and even the more modest requirements of the TGV (considered in the next chapter) was out of the question. This reflected both the economics of railway operations in Britain (as discussed in Chapter 2) and the political reality that funds for British Rail were very restricted compared to other transport projects. From its beginning in 1967 to 1982, the APT project cost £43m, under £3m a year. Over the same period expenditure on new motorways and other trunk roads was £5,300m (averaging over £400m a year, from 1968/69 to 1981/82) and the Concorde project alone cost £2,000m to develop from 1962 to 1976 (£140m a year). The Paris–Lyon TVG line and stock cost about £1,000m. (These sums are at the price levels of the time. At 1985 prices, the APT's development costs would be around £100m.)

A research team for the APT had to be built up from scratch and laboratories and test rigs built at the RTC. The initial project team consisted of about thirty people and to begin with it was hoped to contract Hawker Siddeley, a large private company with considerable railway engineering expertise, to help out with the design concept of APT. In practice it was found to be impractical to involve outside people at such an early stage of project definition and the contract was cancelled. The core design team was built up within BR Research and contractors were involved once design specifications had reached a sufficiently advanced stage. In conjunction with the establishment of an APT project team, research and testing facilities were built. At the RTC test rigs and a special test hall were built and a disused line near Derby was converted into a test track.

So, in the initial stages of the project, the APT faced two major constraints. One was simply that of getting a new research team and research resources together. Money for this was freely available. The second constraint did relate to money, in that the train that the research team were developing with a target operational speed of 250 km/h (155 m.p.h.) would have to use the same signals as trains running at under two-thirds its speed and to operate on the existing rail network with its wide mix of operational speeds and traffic. The alignment of track, particularly the very winding, steep West Coast main line, imposed further constraints on the design brief of the APT. Although taking bends at high speeds did not present a technical or safety problem, it would be uncomfortable for passengers. Finally, in order to maintain the commercial attraction of the APT, speed could not simply be bought by high fuel consumption.

The research and development programme for the APT, approved in January 1969, had the following objectives for the train:

1. Maximum speed 50 per cent higher than existing trains.
2. Curving speeds 40 per cent higher than existing trains.
3. To run on existing tracks within the limits of existing signalling.
4. To maintain standards of passenger comfort at the higher speeds.
5. To be efficient in energy consumption.
6. To generate low community noise levels.
7. To maintain existing levels of track maintenance.
8. To achieve a similar cost per seat-kilometre to existing trains.

These objectives determined the main innovations that the APT had to incorporate, which together represented the most radical jump in rail technology ever attempted. The main innovations incorporated into the APT were as follows:

1. Bogie and suspension design to prevent 'hunting'

The impetus for this innovation was the design speed of 250 km/h (50 per cent higher than existing trains), but to a large extent the design speed arose out of the development of this suspension design in the first place. High speed was assumed to be commercially desirable and this was considered the fastest speed that could initially be achieved.

2. High performance brakes to stop train from 250 km/h, with the use of existing signalling

The impetus for this was really the general approach of the APT project which was intended to avoid the need for expensive infrastructure when market conditions and political attitudes made this unattainable.

3. Tilting body

This was to maintain passenger comfort while train took curves at high speed. The impetus for this was the use of existing track combined with the design speed. However, the ultimate impetus was part political and part economic. Not only was the total rebuilding of rail network uneconomic, but political constraints on railway finances meant that even limited investment in new lines was impossible.

4. Lightweight construction and aerodynamic design

The impetus for this was primarily economic.

5. Powerful, lightweight engines

The choice of motive power was basically technical, but was derived from the design speed of the train and also influenced by the need to keep overall weight low.

6. Low unsprung mass

The impetus was primarily economic in order to reduce

track wear and maintenance, which BR Research had discovered was largely a function of that part of the total mass of the train not cushioned by the suspension (i.e. the wheels, axles, brakes and anything mounted beneath the suspension springs).

Overall, the design brief for the APT was for a 250 km/h train that would broadly have the same operational economics as a 160 km/h train in terms of cost per seat/km. This measure was used because it was recognized that items such as fuel cost were bound to be higher at 250 km/h than at 160 km/h, but that the faster speed would permit the train to be used more intensively, thus compensating for the additional costs of speed.

Following the approval of the design brief, the APT project went through two distinct, though overlapping, stages: an *experimental* R & D phase and a *prototype* design/development phase. An overview of the whole project is shown in Figure 21 on p. 118.

Development of the experimental APT-E

The first stage of the APT project saw the assembling of a project team and resources within BR Research to establish the feasibility of the APT concept. The basic design concept and brief, as outlined above, was used to build an experimental gas turbine powered train, the APT-E, which was completed in 1972 (See Figure 11). Before this, a great deal of design and simulation work was undertaken on both the train design and its individual components. This involved the use of other full-scale vehicles, including the APT-POP train and the APT 'Hastings' Coach, which were built to test particular items of equipment such as the suspension, bogies and tilt mechanism (Figure 12). This was all in order to put together, not a passenger train prototype, but an experimental train, the systems of which could form the basis of a passenger train design.

Figure 11. The APT-E
The ATP-E was a small 'testbed' train to test out the key innovations required for a fast tilting train
Source: British Rail

Initial progress on the APT project was rapid. It took only three years to build up a design team from scratch, to construct and equip laboratories and test track and to design and build a totally new train.* The first test run of the APT-E took place in July 1972, but an industrial dispute over the proposed single driver operation of this train and the diesel High Speed Train led to the suspension of the test programme until September 1973.

Once the trials began again, the APT-E performed promisingly. In August 1975, while on a run between Swindon and Reading, a speed of 245.4 km/h (152.4 m.p.h.) was reached, a British speed record. In October 1975, the APT-E completed a run between London and Leicester at an average of 163 km/m (101.5 m.p.h.). Having completed 37,908 km (23,560 miles) in test runs,

* A full account of the complex research and development programme involved in building the APT-E is to be found in a book by one of the engineers involved, Hugh Williams (Williams, 1985).

Figure 12. The APT-POP Train
This was built to test bogies and suspension for the APT design. To the rear can be seen the APT Hastings Coach. This was adapted from a carriage used on the London–Hastings route. Its narrow body width enabled it to be used for tilt suspension tests of up to 6°

the APT-E was retired to the National Railway Museum in York in April 1976.

Development of the prototype APT-P

BR Research had proved the basic viability of the APT concept; it was possible to build a high-performance train to run on existing track. As the project had progressed, the interest of the passenger business side of BR had increased. In 1971 the British Railways Board undertook a review, *Inter-city Passenger Business: A Strategy for High Speed*, which advocated the commercial development of the APT together with the more rapidly developed diesel High Speed Train. But it still took eighteen months to finally decide how to proceed with the APT project. In 1973 a development brief from Passenger Business management required the APT to be developed as a large twelve-to-fourteen carriage train. Otherwise its design concept, in terms of speed, performance and cost, was essentially as conceived by BR Research. Having established the technical viability of the APT concept, the project now moved to the 'Design—Build—Test' phase for it to be developed into a practical train. With this, responsibility for the project was passed from BR Research to the Chief Mechanical and Electrical Engineer's traction and rolling stock design department. This included the transfer of the APT project team. Authority for the construction of three APT-P prototypes was given in September 1974. Two years of design work went into turning the experimental APT-E into the prototype APT-Ps and the following section summarizes the design requirements and development of the main innovations and design features of these prototypes.

Innovations of the APT-P

The preceding section identified five basic innovative features of the APT design. These are now examined in more detail in terms of the design brief that was established in 1973 for the prototype APT-P trains. An examination is also made of the method of traction that was finally adopted, which led to a need for further design innovations.

Bogies and brakes

The key bogie and suspension design elements to eliminate 'hunting' paved the way for the development of fast trains in Britain. These began to be applied quite generally on most types of new rail vehicles from the late 1960s as new stock was designed and built. In particular, the suspension and bogie design work for the APT made possible the rapid development of the considerably more conventional High Speed Train, which is the subject of Chapter 6 of this book.

The brakes of the APT, however, were unique to this train. They were hydrokinetic (or water turbine) brakes, mounted inside hollow wheel axles. These consist of a double chamber, half of which is static and half of which rotates. Both chambers house turbine blades within them. When the train is moving, these chambers are empty. To apply the brakes, the chambers are simply filled with water. The water in the rotating chamber churning against the static turbine blades in the fixed chamber slows down the rotating chamber which is linked to the wheels through the hollow axle. Braking energy is converted into heat and the hot water is transferred to a radiator on the body of the train to be cooled. The reason for choosing this braking system was primarily that it was very light (110 kg) for the performance that it can achieve. Brakes have to be located on the unsprung part of the train, below the suspension system, and for reasons of track damage and wear (considered below) it was vital to keep the unsprung mass of the APT very low. Disc brakes, as used on the High Speed Train and other locomotives and rolling stock, are much heavier and it was doubted whether they were viable for braking from 250 km/h. However, hydrokinetic brakes, since they depend on the churning of water, become inefficient at low speeds and so a secondary set of light-duty tread brakes were fitted to gradually take over the braking effect at speeds under 80 km/h.

Track damage

The unsprung mass of the train is of particular relevance as this is a major determinant of the wear, tear and damage caused by a train on the track, rather than just the overall mass of the train. This was another research finding to arise from the fundamental work carried out by BR Research in the 1960s and from the experience of running 160 km/h services. It was decided that the track forces produced by the existing Deltic locomotive at 160 km/h should not be exceeded by any future trains, even at higher speeds. Theoretical work, together with tests on the West Coast line, showed that, to achieve this objective for operations at 200 km/h, the unsprung mass of any pair of wheels must not exceed 2.5 tonnes. For the 250 km/h design speed of the APT an unsprung mass of 2.2 tonnes per wheel-set was required. Hence, one criterion by which many rail innovations are now judged is their effect on unsprung mass. For the APT this was particularly important as it would be taking curves at up to 250 km/h, which on a conventionally designed train would cause an incredible amount of track damage and would probably not be safe (as high-speed runs using ordinary locomotives in France in the 1950s had shown). Not only the brakes of the APT, but a number of other innovations were chosen or modified owing to their effect on unsprung mass.

The main savings in unsprung mass came from the mounting of the electric traction motors inside the sprung body of the power car driving lightweight gearboxes on the bogies connected via flexible couplings to the axles of the driving wheels. The use of articulated bogies (considered below) also reduced unsprung weight and the tilting of the train into curves helped a little by keeping the train's weight centrally on the suspension. The end result was that the APT had an overall unsprung mass of only 1,500 kg (compared to 3,300 kg for a Deltic hauled train), and in tests an APT travelling on straight track at 200 km/h exerted only half the track forces of a Class 87 locomotive hauled train at 160 km/h and exerted the same track forces as the Class 87 on curves.

Signalling and operational speeds

The maximum operational speed of the APT, 250 km/h (155 m.p.h.), was the figure estimated by its designers to be the highest speed technically and economically feasible for the project. In the early years of APT, even faster speeds of up to 300 km/h were mentioned, but for the APT-P prototypes, 250 km/h became the target. This had major implications for a number of innovations in the train's design, but for signalling the problems were more acute. The 'block signalling' method on Britain's railways divides all track into equal sections. The signals on these blocks have four 'aspects': green, double amber, single amber and red. This sequence gives a braking distance of 2,050m (6,728 ft) (see Figure 13), which is perfectly adequate for trains using conventional tread brakes at up to 160 km/h and for disc or hydrokinetic brakes at up to 200 km/h. Above this speed it is simply not possible to stop a train in 2,050 m as the adhesion of steel wheels to steel rails is insufficient. The train would simply slide along the track.

To operate the APT safely, a 'fifth aspect' would have to be provided to give a driver a warning to slow down to 200 km/h where conventional signals showed green one block before a double amber. This gives a total stopping distance of 3,075m. Totally resignalling the routes to be operated by the APT by installing a 'fifth aspect' to the signalling blocks was seriously considered by British Rail, but the costs were seen as prohibitive if conventional signals were to be used. In addition, there were doubts as to whether it would be safe to add a fifth aspect to visual signals, since in marginal situations it could confuse drivers.

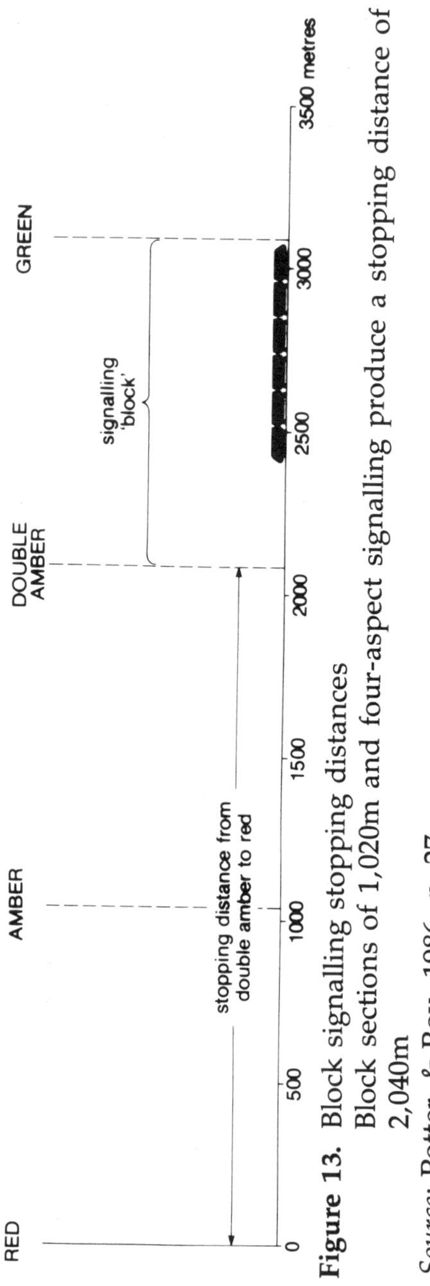

Figure 13. Block signalling stopping distances
Block sections of 1,020m and four-aspect signalling produce a stopping distance of 2,040m

Source: Potter & Roy, 1986, p. 27

Another approach would have been to retain the four aspect signals but to lengthen the block sections in order to increase braking distances. This was more or less what had been done with the French TGV, but in this case longer braking distances on the new Paris–Lyon line were combined with an electronic signalling system. But to do this on Britain's existing lines would be prohibitively expensive. All signals would have to be moved and all the control methods and signal boxes modified. This would, of course, only hold for the APT, since the vast majority of trains, which run at lower speeds, have no need for such signalling. On a purpose-built track, on to which all fast trains are concentrated, as is the case with the TGV, this is economically justifiable. In the British context, it is not.

Quite remarkably, this problem was simply shelved. Except for special tests, when the track was cleared ahead of it, the APT was restricted to 200 km/h. The 'fifth aspect' question was put to one side.

Tilting body

The tilting body is perhaps the one feature most associated with the APT. Many of Britain's most densely-used rail routes, particularly the electrified West Coast main line, have a high proportion of curved track. Overall, 50 per cent of British Rail's track is curved. Speeds are restricted around curves depending on the curve radius and the banking of the track around the curve (called 'cant'). Passenger tests and operational experience had established that the maximum centrifugal force considered acceptable was about 0.07 g, equivalent to a 'cant deficiency' of 4.25° (i.e. the cant of the track would have to be built up by another 4.25° for the passengers not to feel any centrifugal force at all—see Figure 14). If lateral forces were any higher, walking along a train would become difficult, objects would slide off tables and drinks would

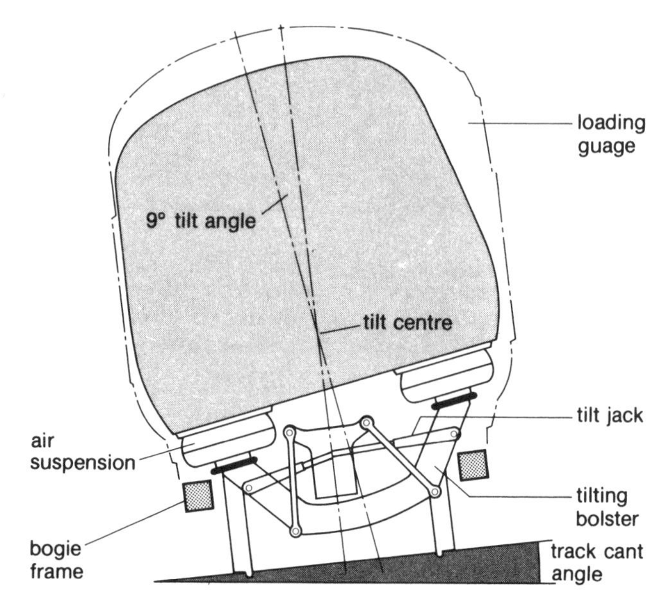

Figure 14. Loading gauge and tilting profile of the ATP-P With the track built up to a maximum 6° cant, and the body of the train tilting at 9°, the APT-P was able to tilt a total of 15° into curves. However, the top of the profile had to be narrowed to keep within the loading gauge.

Source: Potter & Roy, 1986, p. 28

be spilt. Canting of the track is limited to around 6°, as any steeper angle, although fine for passengers in fast trains, would be uncomfortable for passengers in slow trains or on a train that stops on the curve.

Slowing down and accelerating away from curves would seriously reduce the average speed of any fast train. Hence, being able to tilt was crucial to operating on existing curved track at above 160 km/h. By being able to tilt at up to 9°, the APT was designed to take curves at up to 40 per cent faster than conventional trains and give a more comfortable ride to passengers by eliminating all centrifugal forces.

The research work done into passenger comfort on curves was old and relatively crude. It had been undertaken in the late 1940s using steam trains on Welsh mountain railways. This suggested a maximum cant deficiency of 3°, which was raised to 4° when the use of continuously welded track improved ride quality such that a little more lateral g-force was considered acceptable. No new passenger tests related to cant deficiency in fact took place until 1983. The tilt was so crucial to the APT concept that further passenger tests were considered unnecessary.

Experiments with tilting trains had taken place since the 1930s. The earliest method was the pendular design, whereby the carriage was simply suspended on a pivot point and was free to swing into curves. This was tried in the United States in 1938–41 and in France in 1947–56. Later two designs of train entered passenger service using the passive pendular method. The first was the United Aircraft 'Turbotrain' in the United States in 1969, which also saw service in Canada at a later date, followed by the Japan National Railways pendular multiple unit train on the Odakyu line in 1973. A pendular design also saw service in Sweden. The passive pendular concept was not used for the APT for two major reasons. Firstly, the 'loading gauge' on British Rail is very tight compared to most other railway networks. The loading gauge governs the maximum width and height of a train (Figure 14 shows the C1 Loading Gauge). If a train exceeds the gauge size it will be in danger of brushing or hitting other trains, bridges and trackside equipment.

The pendular tilting method requires a high pivot point in order to get a carriage to swing into a tilt by its own momentum. Unless a carriage were very narrow indeed, this would take it outside British Rail's loading gauge. As such, the design of the APT adopted a low pivot point, in order to have a profile only very slightly narrower than an ordinary train, yet was able to tilt by 9°. This keeps the APT design within the loading gauge, but with the low pivot point virtually at the centre of mass, there is no pendular effect. The tilting method has to be powered. At the same time as the APT, active, powered, tilt mechanisms were under development for rail coach designs in Canada, Spain, Switzerland, Italy and Sweden.

The tilting of the APT had a major influence on a variety of design features of the train. The top of the coach is somewhat narrower than a non-tilting train and its interior design seeks to disguise this. Also links between the tilted and untilted parts of the APT need to be flexible or have a compensatory action. (E.g the overhead power collecting pantograph.) Also the unique profile of the APT means that special washing facilities had to be constructed as existing ones were designed for wider trains.

The tilting mechanism for the APT involves hydraulic jacks mounted on the bogies, which tilt each part of the train in response to signals from an electronic sensor measuring lateral acceleration. After initial tests on the APT-P prototypes, the sensors were set on the coach in front of the one to be tilted in order that response to curves would be faster.

Lightweight construction and streamlining

The lightweight, streamlined construction of the APT is a feature common to all fast trains. The relative importance of weight and aerodynamic drag varies with speed, drag increasing in proportion to the square of speed. As in the case of the development of the A4 Pacific steam locomotive in the 1930s, wind tunnel tests were undertaken to establish the optimal shape for the front of the APT. Early sketches and models featured a crudely streamlined bulbous nose, vaguely similar to the Shinkansen, but the end result was a 'wedge' shape, essentially the same as that adopted by Gresley for the A4 Pacific (compare Figures 2 and 15). Other basic aerodynamic features of the APT

Figure 15. The APT-P undergoing commissioning trial on the West Coast main line
Source: British Rail

harked back to Gresley's design, such as the streamlining of the bottom of carriages and vehicle articulation.

The lightweight construction of the APT particularly saves fuel at low speeds and reduces the kinetic energy to be dissipated during braking. It is also very necessary in order to reduce axle loading on the articulated bogies and to reduce track damage of high-speed operations. Streamlining and a variety of techniques to save weight were therefore used in order to reduce the APT's fuel consumption. The end result is that, at comparable speeds to existing trains, the APT uses a third less fuel per passenger. But at an operational speed of 260 km.p.h. it uses over a third more fuel per passenger. This gives an idea of the fuel penalty of high-speed rail operations and the resultant need for streamlining and weight savings. A major saving in weight was achieved by the use of aluminium alloy for the passenger coaches, although steel was retained for the power cars because of the additional stiffness required. To achieve such load-bearing in aluminium would require very thick sections, possibly even weighing more than steel.

For the APT-E, aircraft construction techniques were used, with aluminium sheeting riveted over the coach frames. Although an acceptable method of building an experimental train, this technique involved a lot of manpower and time and, especially given the high cost of aluminium, was too expensive for a commercial train. Hence, for the APT-Ps, a new manufacturing process was developed in order to overcome the high cost of materials and existing manufacturing methods. This process was developed jointly with the Swiss aluminium company, Alusuisse, and involved the use of commercially available wide aluminium extrusions which were seam-welded together automatically to run the length of the vehicle. This gave a body shell of a uniform cross-section, reinforced near the ends and stiffened by triangulated hollows formed in the extrusions or by additional frames where necessary.

This innovation in railway vehicle construction was successful in that it more than compensated for the increased cost of materials, and the design has since been sold to a number of railway coach manufacturers. This design innovation resulted in the APT coaches weighing 40 per cent less than conventional steel coaches and very much aided the other major weight-saving feature of the APT, the use of vehicle articulation. Adjacent passenger coaches shared a bogie, hence resulting in a third fewer bogies being needed than for a conventional train of the same size. Articulation also reduced air-flow drag. One unusual weight-saving innovation was the use of chemical toilets in the APT. These used a recirculating sterilizing fluid for flushing and saved 0.75 tonnes per coach, compared to a conventional water-flush toilet.

Motive power and traction

When the APT project was moving towards the construction of the APT-E in the late 1960s, it was envisaged that two types of motive power would be used: electric traction and, for non-electrified routes, gas turbines. Diesel power was not considered to be viable since, above 200 km/h, the ratio between power and mass becomes very poor. In fact gas turbines were chosen for the APT-E, which was completed in 1972, partly because the non-electrified Newcastle to Bristol route (see Figure 5) was in mind for the first application of the APT. This is a very slow inter-city line with a high proportion of curved track. Being able to curve more quickly on this route would have resulted in a major reduction in journey time. The APT-E had two power cars, each with five prototype Leyland gas turbines. At this time the Leyland company (now part of Rover) was developing these light and powerful gas turbines for use in heavy lorries. However, the 1971 review of the APT decided that the first application for service trains would be on the electrified West Coast main line and not the cross country Newcastle–Bristol route, and hence work shifted towards electric traction for the APT-P prototypes. It was still intended to produce a gas turbine version, with two electric and four gas turbine prototypes planned, but the gas turbine option ceased when, following the 1973 'fuel crisis' and the sharp rise in the cost of oil, Leyland abandoned the development of these engines.

The 1973 passenger specification for the APT-P was for a large 534-seat, twelve-coach train (with a potential stretch to a fourteen-coach, 653-seat train—the maximum that station platforms could accommodate). This required the use of two power cars as it was simply not possible to build a single power car large enough to accommodate a 6,000 kW engine. Even if there were one 6,000 kW power car, it could not deliver that power through only four axles, because there would be insufficient friction force available and the wheels would simply spin and slip before full power output was reached. To use 6,000 kW, two power cars and eight powered axles were needed.

This, however, presented a new design problem, which was where to place the two power cars within the train. The overhead wires on the West Coast main line had been installed assuming a maximum operational speed of 160 km/h using conventional locomotives with a single power collecting pantograph. To have two 3,000 kW (4,000 hp) power cars at each end of an electric train would require two pantographs. At high speeds the front pantograph would disturb the overhead wire, causing periodic and random loss of contact for the rear pantograph. This would cause arcing and damage to the current collector shoe and driving motors. A different design of overhead wiring could overcome this, but for reasons of cost this was out of the question. The option of having a single pantograph with a high voltage power cable running along the train to the second power car was considered unsafe because of tilt movements. The non-tilting TGV has such a cable running along its roof.

The only practical solution was to locate the two power cars together so that they could be served by a single pantograph. This had to be either at one end of the train or in the centre. The option to have both power cars at one end of the train was ruled out because of problems that would be created when the train was being pushed from the rear rather than pulled. The thrust against the adjacent carriage would be so great that it would buckle the suspensions against the bump stops. Thus the power cars were positioned in the centre of the train. One implication of this design was that the train was effectively split in two. There is a corridor running through the APT power cars, but it can only be used by staff as it is too narrow and noisy for passenger use. Thus, each half of the train has to have its own staff and catering facilities. But for a twelve-to-fourteen carriage train it is likely that this would have been necessary anyway.

APT innovations: technology push or market pull?

The Advanced Passenger Train was born out of a reorientation of research in British Rail from that of a 'defensive' strategy (focusing on short-term technical improvements to existing designs) to a scientific-based 'offensive' approach in which BR Research would be the initiators of a project. So, although the APT was conceived at a time when faster trains were having a major impact on BR's passenger traffic, it was basically a product of BR's Research Department. They believed that a proposal for a very fast train would be favourably received by the commercial management within BR, which indeed it was. The basic concept was broadly technically led—the design speed was well above anything that the marketing management of BR could evaluate and their eventual 'design brief' was little more than a modified version of the existing APT specifications!

This raises an important question as to the initial stimuli for innovation. In an economically competitive environment such as passenger transport it might be expected that market needs and commercial judgements would be the way by which the need to innovate would be identified. If commercial management notes that fast trains, for instance, attract passengers, market research is then undertaken to provide a specification for engineers to design and build a particular type of train. This 'market pull' approach to innovation, found in its extreme form in careful market research as championed by Levit (1960), seems logical and sensible. But, in practice, a totally market-led innovation process is rare. What happened with the APT (and with most other train projects) is that the engineers responded to their knowledge of the rail market, and what was happening in other countries like Japan, Italy and France, produced an innovative design and then sought support from the commercial and operational side of BR to have it accepted. The change of role for research in BR and the establishment of the Railway Technical Centre only really represented a shift in power *within* engineering. As the historical case study of fast steam trains and attitudes towards electrification in Chapter 2 showed, the engineers have always had the upper hand in initiating rail innovation.

This is not to say that commercial and economic considerations were ignored in drawing up the design of the APT. Far from it; but it was the scientists and engineers who made commercial judgements in formulating the initial design specifications and not the commercial operational side of British Rail. Whether this is necessarily a bad thing is another question. Commercial management may be unaware of what is technically possible, as was clearly the case with the APT, and as such it is very hard for radical innovations to be market-led. Conversely, by making commercial judgements themselves, engineers in devising an innovation may well produce a design concept which, though attractive to the commercial management side of the railways, may incorporate some incorrect commercial assumptions that are hard to rectify.

This raises an important issue as to the relative role of 'market pull' or 'technology push' in determining what innovations may be developed. Bennett & Cooper (1979, reprinted in Roy & Wield, 1986) summarized the situation as follows:

> The marketing concept suggests that, in their new product efforts, firms must be customer oriented . . . buyers' needs and wants should be identified, qualified and quantified as part of product idea conception . . . . But the evidence over the years suggests otherwise. In fact, many of the great product innovations throughout

history have been the result of a technological breakthrough, a laboratory discovery, or an invention, with only a vague notion of a market need in mind. Often these 'great ideas' originated from men and women far removed from customers . . . who . . . in the normal pursuit of scientific knowledge, gave the world the telephone, the phonograph, the electric light, and in more recent times, the laser, xerography, instant photography and the transistor. In contrast, worshippers of the marketing concept have bestowed upon mankind such products as new-fangled potato chips, feminine hygiene deodorant and the pet rock.

Lorenz (1983) neatly sums up the tension between market pull and technology push approaches to innovation:

> The problem . . . with many studies even now is that they simply measure what the consumer knows he or she wants. This may be fine when a company is deciding whether to change the colour of an existing product, to improve its reliability, or to add a new feature. But when the firm is dealing with an entirely new product concept, it is a completely inappropriate form of research.

When an innovation produces a product that is significantly different from consumers' experiences, the results of market research are bound to be imperfect. An entirely market pull approach to innovation cannot cope with technical breakthroughs. In the case of the APT there was a blend of technology push and market pull. The initial APT proposal was certainly the result of a technical breakthrough, but it would probably have been rejected if the British Railways Board had not been keen to have a fast train to compete with road and air transport. It may have been impossible to evaluate the market impact of the APT, but clearly it was something that had commercial appeal. Once the project got going, marketing inputs became stronger, with the APT designers having to cope with several changes of specification due to changing market requirements. Certainly, after the initial push from technology, pulls from the market took over with a vengeance!

Overall, as Freeman (1982) notes, whether an innovation is commercially or technically led, what is perhaps the more crucial element for success is a good working relationship between those designing a product and those marketing it. This relationship did play a crucial part in the eventual outcome of the APT project, but in the meantime relationships *within* the research and development structure of British Rail were destined to dominate the project.

5 The TGV

Britain was not alone in conducting fundamental research into the suspension systems of high-speed passenger trains. Following their experimental high-speed runs in the mid-1950s, which more than anything showed how dangerous it would be to run conventionally designed trains at high speeds, the French national railways, SNCF, undertook a programme of research and experimentation in fast passenger train design. This involved a double strategy: firstly there were 'evolutionary' type improvements to existing locomotive and carriage designs, and by the late 1960s this had resulted in SNCF having in service a number of 200 km/h locomotive-hauled trains. However, SNCF were convinced that 200 km/h represented the practical limit to conventional railway technology. The second part of their strategy involved similar 'scientific' research techniques to those being undertaken by BR Research at the Derby Railway Technical Centre. Although some of the resultant design features have their parallels in the APT, there are striking contrasts between France's approach to high-speed passenger trains and that of Britain.

The French experiments in high-speed running, begun in the mid-1950s, continued into the 1960s with the intention of developing lightweight, powerful trains that could safely run on conventionally built railway tracks. As in Britain, keeping unsprung mass low and a suspension design to raise critical speeds and so eliminate 'hunting' formed the focus of the research work. In 1967 a railcar was modified to accept an aircraft gas turbine engine. This RTG railcar reached 252 km/h during trials, and subsequently fifty-three 200 km/h gas turbine railcars were put into service to gain practical high-speed operational experience.

As was the case with the APT-E, the next stage was the development of an experimental high-speed train to test all the practical systems required. This was the prototype Train à Grande Vitesse (TGV 001). Like the APT-E, TGV 001 was a double power car, highly streamlined, lightweight gas turbine train. As was the case with the APT, there was a shift to electric traction following the 1973 energy crisis. But there the similarity between the two projects really ends, for both the design requirements and technology incorporated into the TGV and APT display a markedly different approach.

Design features of the TGV

The design speed of the TGV was much higher than for the APT. The prototype was intended to undertake experimental trials in the 250–300 km/h range, and in December 1972 reached 318 km/h. The production version was designed to operate at 260 km/h (later raised to 270 km/h), but incorporated a 300 km/h capability. In 1981, a production TGV ran at 380 km/h. However, despite this higher performance, the design on the TGV was technologically much simpler than the APT.

The design features closest to the APT were general aerodynamics (both had a wedge-shaped cab), plus measures to reduce the unsprung weight of the train in order to reduce wear and tear on the track. This involved the mounting of traction motors in the body of the train connected to the wheel-set via a cardan shaft, very much the same approach as adopted in the APT. Another similarity with the APT was the use of articulated bogies to save weight, but the general suspension design was somewhat simpler. This involved the development of an existing bogie design, modified to include dampers to reduce hunting vibrations. However, crucial to this design was the use of low cone-angled wheels, so that the technology was really a combination of traditional 'evolutionary' and 'scientific' techniques.

What made this somewhat simpler suspension design possible was the fact that the TGV would only run above 200 km/h on purpose-built track with gentle curves. The development of the high-performance TGV was entirely within the context of the construction of a major new line, the design of which was integrated with that of the TGV trains to run on it.

The reasons for the construction of the new Paris–Lyon line are considered below, but the fact that the TGV would only run fast on the new line meant that many innovations incorporated into the APT were entirely unnecessary for the TGV. A tilting body was not needed, which made it possible to run a 25 kW cable along the roof of the train connecting one pantograph to the two power cars at either end. The overhead wires were specifically designed for power collection at speed, but even so the pantograph design did present one of the major developmental problems of the project. Also, because of the new line, conventional brakes could be used; braking distances were simply lengthened. The only major innovation incorporated in the construction of the new Paris–Lyon line was the use of electronic signalling, using track circuits which display signalling information on instruments in the driving cab. The reason for this was simply the difficulty of drivers seeing visual signals when travelling at up to 300 km/h.

Streamlining and a lightweight construction were clearly important, but basically the suspension system represented the only crucial innovation necessary for the TGV. Even so, as noted above, this was less innovative than the suspension on the APT as the use of specially aligned track allowed low cone angled wheels to be used. Overall, the crucial fact that the TGV was only planned to operate at high speeds on purpose-built track meant that this far less innovative train was capable of exceeding the performance of British Rail's highly innovative APT.

The design image of the TGV was considered to be crucial and SNCF were looking to this train to enhance the whole image of rail travel in France. For this reason, they employed an external design consultant who worked together with the engineers and managers on the TGV project team, and who had total control (within technical limitations) for all design elements of the train. This varied from the whole external train profile, through the colour (bright orange!), down to the small details such as door handles.

The project was managed using a 'matrix'- type management structure. Management structures will be discussed further with respect to the APT, but basically, project control for the TGV involved two distinct stages, mirroring those of the APT. The first, the research stage, involved, as with the APT, a tight research project team. The second stage is where contrasts with the APT emerge, for although for its design and development there was still a TGV project team, the form this team took was significantly different to that of the APT. Rather than it being the existing research team moved into development, it consisted of people who were drawn from the different departments of SNCF, from industry and from private consultants, as those skills were needed. This flexible 'matrix' management structure was the normal method used by SNCF for major new projects, which contrasts with how British Rail had to develop new management structures for the APT project itself.

The new Paris–Sud-Est TGV line

It was in 1969 that SNCF submitted plans for a new line between Paris and Lyon to the French government. This was the busiest rail route in France, and was heavily congested. In particular, there was a 107 km (67 miles) bottleneck to the north of Dijon where the line narrowed from four to two tracks. SNCF examined a range of alternatives to tackle this problem, concluding that the construction of an entirely new line between the outskirts of Paris and Lyon presented the best solution. This was the most expensive of the options considered but, if built as a specialist high-speed line, the revenue-generating potential tipped the balance in its favour.

The key to this revenue-generating potential was that the new line, although only 415 km (258 miles) long, would act as a fast trunk route for trains operating on 1,625 km (1,010 miles) of other lines linking in with it. This meant that a limited investment in this new line would substantially upgrade a large part of France's rail network. It was the traffic-generating potential of this integrated package that gave the plan for the new Paris–Sud-Est line the advantage over the piecemeal upgrading of the old line. The French government accepted this argument and in 1971 gave the go-ahead for the project. Following a public enquiry in 1975, construction of the new line began in 1976.

The reason for the French needing additional track capacity stems back to the origins of inter-city railways in France. Although the railways were built by private enterprise, this occured within a very strong structure of state control. Successive governments, from the mid-nineteenth century, had ensured that unnecessary duplication of lines did not occur and that the private lines meshed together to form a national network. Hence, on each major route there was only one line. By the 1960s, the Paris–Lyon line was nearing its capacity.

Exactly the reverse situation occurred in Britain where state control over railway development was very weak. In consequence several different companies built competing lines along the same route and by the time the railways were nationalized there was a considerable over-capacity. As such, except on a few suburban lines, capacity is not a problem on Britain's railways.

Because the new line was for the exclusive use of TGV trains it was possible to avoid the need for tunnels by using steep gradients. The southern part of the Paris–Sud-Est (TGV–PSE) line was opened to TGV passenger services in 1981 and the northern link to Paris opened in 1983. The Paris–Lyon journey was reduced from 3h 48 min to exactly two hours and even on routes involving substantial running over existing tracks, major time savings have been achieved (e.g. the Paris–Geneva time was reduced from 6h 14 min to 3h 30 min). Today the 270 km/h TGV is the fastest passenger train in the world (Figure 16).

Figure 16. A TGV-PSE train at Le Creusot, a new station on the Paris–Lyon line

The financial performance of the Paris–Sud-Est TGV

The total cost of developing the trains and building the new TGV-PSE line was just under £1,000 m. Traffic growth has been substantial. As Roberts & Woolmer (1984, p. 72) noted:

> In the first year of operation the TGV attracted on average an extra 5,300 passengers a day. This represented 35% of the TGV passenger business and an increase of 14% on all rail traffic on the Sud-Est routes . . . . After 18 months of operations the TGV carried 10 million passengers, reaching 14 million in September 1983.

Overall, as noted by Walrave (1985), rail traffic on both the TGV and existing Paris–Sud-Est routes grew from 12.2 million passengers in 1980 to 18.4 million in 1984, with TGV-PSE services representing 17 per cent of SNCF's total traffic.

In assessing the financial performance of the TGV-PSE investment in track and stock, account has to be taken of the effects on the Paris–Sud-Est network as a whole, for it was as part of an integrated project that this investment was authorized. It is deceptive to evaluate the TGV investment in terms of the total passenger revenue that the system now earns simply because most of the traffic on the TGV-PSE route is merely a transfer from the old line. A more valid evaluation would be to examine the additional revenue generated by the investment in the new line and the TGV rolling-stock. According to Walrave's figures, by 1984 directly generated additional revenue was running at 1,362 m frs per annum. This may be compared with the annual cost for stock depreciation and the financial charges of building the new line of 1,145 m frs, yielding a net surplus of 217 m frs. Calculated over a twenty-year period, an internal rate of return of 20 per cent is estimated for the toal TGV-PSE line investment.

Towards a TGV network

The success of the TGV Sud-Est paved the way for a second TGV line, the TGV Atlantique, and now SNCF are confidently looking towards a TGV Nord, linking Paris with Brussels, Amsterdam, Cologne and to the Channel Tunnel. There are even plans for a fourth line to southern Germany. What started off as a specific project has now developed into a plan for a core network of 300 km/h high-speed lines which, well integrated with existing lines, will transform inter-city rail operations over much of Northern Europe.

It was in 1984, just a year after the full TGV Sud-Est service began, that authorization was given for the construction of an Atlantic Coast TGV line. The rationale is much as for the Paris–Sud-Est route, with a new line dividing into two which then by-pass capacity bottlenecks at Le Mans and Tours. The 280 km (174 miles) of new line are estimated to cost 9,400 m frs (£850 m), but because they link into a series of routes to cities along France's Atlantic coast, this investment will have a major impact on the upgrading of more than 1,600 km (1,000 miles) of track. The electrification of two connecting lines to Le Croisic and Angers and the upgrading of other lines to 200 km/h operations also form part of the TGV Atlantique project. This network already carries 15.5 million passengers a year and the investment study undertaken for the project envisages this rising to 18.8 million, resulting in a 10 per cent internal rate of return.

Technically, the TGV Atlantique (TGV-A) trains will be similar to those used on the Sud-Est route, but as the track gradients will not be as great, longer trains will be used, seating five hundred people. The main technical change is in the raising of operational speeds to 300 km/h, a result of improvements to the engine design, involving the use of synchronous motor traction. Given that the new line will represent a small proportion of most journeys, 300 km/h as opposed to 270 km/h running is of greater publicity than practical value. The 300 km/h running will be of more commercial importance on the proposed TGV Nord routes to Belgium, the Netherlands, north Germany and, via the Channel Tunnel, to Britain. The development of a 300 km/h capability for the Atlantique line certainly places the French railway industry in a strong bargaining position relative to other European nations.

The largest modification to the TGV-PSE design is not technical but in terms of interior design. The TGV-PSE trains had a somewhat austere and cramped interior with 'airline' style seating in both second and first class. Marketing relied very much on the TGV's speed and premiere image rather than interior comfort. The TGV-A trains will contain a remarkable variety of accommodation, varying from a small conference room for business people in first class through to a compartment in second class with lifting seats designed as a childrens' play area, complete with toys and games! Construction work on the TGV Atlantique began at the beginning of 1985 and the Le Mans and Tours branches of the line are due to open in 1989 and 1990 respectively.

In July 1983 a meeting of the Transport Ministers of France, Belgium and West Germany discussed the concept of a high-speed line between Paris, Brussels and Cologne. This would involve 420 route km (261 miles) of new track at a cost, estimated in early 1986, of £630 m. With 300 km/h running, a Paris–Cologne time of 2h 30 min, averaging 240 km/h (150 m.p.h.) is envisaged.

A preliminary report in late 1984 deemed the project financially sound and detailed studies were authorized. Meanwhile, remarkably rapid progress was made on proposals for the Channel Tunnel, the go-ahead for which

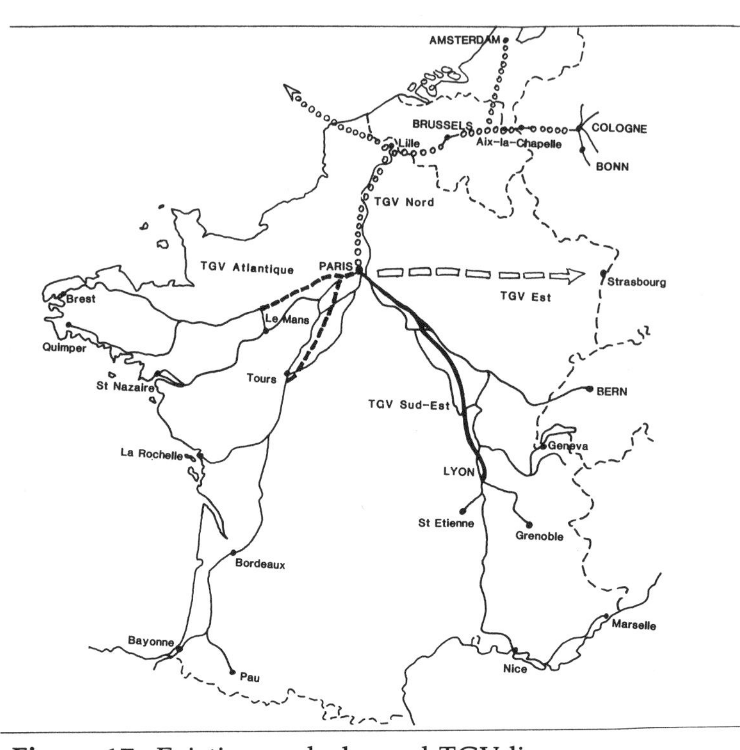

Figure 17. Existing and planned TGV lines
The new Paris Sud-Est TGV line links into a number of existing lines serving Southern France and into Switzerland. Likewise, the TGV Atlantique will consist of two new lines linking into a number of existing routes to France's Atlantic seaboard cities. The TGV Nord moves into the multinational league, the exact route of which is yet to be determined. Only conceptual plans exist as yet for the TGV Est to Strasbourg and Southern Germany

was announced in early 1986. The Channel Tunnel would be served by a spur from the TGV Nord, and with it due to open in 1993, an acceleration of the plans for the TGV Nord seem likely (this is looked at further in Chapter 8).

A second trans-European TGV line is also being studied. In January 1985 a working party was set up to study the concept of a Paris–Strasbourg–Stuttgart line to link in with southern German and Swiss lines and the new German high-speed line between Stuttgart and Mannheim (Figure 17).

Anglo–French contrasts

The TGV represents a very different approach to that of British Rail. Compared to the APT, the TGV is a technically less innovative design, made possible because favourable geographical, historical, economic and political circumstances have made the construction of a limited amount of new purpose-built track a viable proposition. But perhaps what is really crucial and innovative about the TGV concept is the way in which this limited investment in new track is carefully planned in such a way as to revitalize a whole network of routes. Less than 700 km (435 miles) of new track are involved in the TGV-PSE and TGV-A projects, yet together with the existing lines with which they link, they have upgraded a network of lines serving 75 per cent of France's population.

Although Britain has a higher population density than France, there is no obvious route along which a TGV-type line would appear viable. There are no crucial capacity problems and no one route which has the growth potential to justify a totally new line. With demand spread over a number of main lines, the historical, geographical and economic situation of British Rail ruled out a TGV-type approach. Having to accept that any future fast train would have to operate on existing, mixed traffic lines therefore imposed a host of additional constraints on the design of the APT compared with the TGV. The situation in Britain required a more innovative approach. Whether it required the particular mix of innovations that were incorporated into the APT design is another question, (to be addressed in Chapters 7, 8 and 9), but certainly as technically simple approach as was adopted by the French was not possible.

But the organizational, systems approach that was adopted for the TGV is another matter. The comprehensive, integrated investment approach, involving both a financial appraisal and cost-benefit analysis, presents a contrasting picture with Britain. Rail investment in this country is evaluated on a narrowly defined project basis. SNCF's argument that the new TGV-PSE and TGV-A lines, although more expensive than other options, would prove more cost-effective because of their effect on traffic generated by the railway network as a whole would not be accepted as a case for investment by the government for British Rail. Projects have to be justified in terms of their own costs and revenue. Cost-benefit analysis is definitely out of the question. Had SNCF's proposals for the two TGV lines been evaluated using the financial criteria that British Rail is required to use, the proposal for a new line would certainly have been rejected in favour of the cheaper piecemeal upgrading of existing routes.

Add to this the political factors of railways having a low priority in government spending plans and hence low investment levels, coupled with strong government support for rail's competitors, then it becomes clear that the development of fast trains in Britain has taken place under commercial and political conditions that are fairly hostile to successful innovation. With the scope for conventional rail developments so restricted, British Rail has had to follow a more radical line in order to develop fast passenger trains. It was these conditions that both stimulated and constrained the development of the Advanced Passenger Train.

6 The evolutionary alternative

When the Advanced Passenger Train project began in 1967, British Rail envisaged that the first trains would be in passenger service by the mid-1970s. But by 1970 it was becoming clear to the British Railways Board that the innovative nature of the APT was going to require a long development period. It was at this point that the Chief Mechanical and Electrical Engineer's Department (CM & EE) began to canvass for a simple, quickly developed, 200 km/h high-speed diesel train. The claim was that a prototype could be running within two years.

What lay behind this proposal was a total lack of confidence by the traditional railway engineers in the APT concept. Not only was the APT viewed as totally impractical but as a positive hindrance, as no other major passenger train developments were being authorized. The idea was not that BR should be seeking a TGV-type approach, but that the potential now existed, by stretching existing designs and methods, to push operational speeds up to 200 km/h which, for some routes, was entirely feasible. For example, the upgrading of the East Coast Main Line for 'Deltic' operations made it suitable for such treatment.

The reaction of the BR Board was pragmatic. They continued to support the development of the APT, but took the Chief Mechanical and Electrical Engineer at his word and, in August 1970, authorized the development of a single prototype 'high-speed diesel train'. This was coupled with a general review of plans for high-speed operations. This review, entitled *Inter-city Passenger Business: A Strategy for High Speed* (mentioned in Chapter 4, above), was prepared by the Passenger Planning Department of British Rail in consultation with BR Research, the CM & EE and other BR departments. It was the first time that the initiative in planning fast train developments had come from the commercial side of BR rather than from research and engineering. Nevertheless, the two main options under consideration, the APT and HST, had been presented to them by the engineers. Market pull was within the bounds determined by technology push!

A Strategy for High Speed was presented to the British Railways Board in May 1971. It was this review that opted for the electric version of the APT and set the groundwork that led to the 1973 specification for the large twelve-to-fourteen carriage prototypes. But this review also supported the continued development of the High Speed Train (HST). The role of the HST was seen as having a threefold capacity. Firstly its quick development would provide BR with a fast train in advance of the introduction of the APT. Secondly, with the APT shifting to electric traction and with there being a need to replace some diesel trains anyway, there was a requirement for a fast, non-electric passenger train as it was. Finally, the HST was seen as providing an insurance against any delays should there be technical difficulties in introducing the APT.

There were two lines on which the inability of the HST to take curves at high speed did not matter greatly:

> When individual main lines were examined in detail, it quickly became obvious that the former Great Western main line from London to Bristol and South Wales, along with the East Coast main line from London to Edinburgh, were far better suited, because of their lack of serious curves, to high speeds than the remainder.
>
> Quite what prompted Brunel to lay out his line to Bristol in the 1830s with curves of several miles radius (at no small expense to the GWR shareholders) when he could scarcely have envisaged speeds above 100 km/h is

something of a mystery. At all events, it made the going easy when the world's first 200 km/h service of diesel trains commenced operation in October 1976. [Ian Campbell, Vice-Chairman of BR Board, writing in *Railway Gazette* (1980), p. VI.]

The High Speed Train had been sold to the British Railways Board as a rapidly developed, fast (200 km/h), yet 'conventional', train. It was developed at the Railway Technical Centre by the CM & EE Traction and Rolling Stock Design Department, who very much seized the opportunity to demonstrate their capability to produce a high-performance train. Virtually the entire resources of this department were devoted to the design, development and testing of the HST, with the result that the prototype was indeed completed in under two years.

The professional rivalry involved between the designers and engineers in the CM & EE Department and the scientists and research engineers in BR Research at the Railway Technical Centre is typical of the internal policitics of any large organization. The 'traditional/evolutionary' school of empirical engineering within BR was seeking to show the 'radical' theoretical research-based BR Research Division that it did not have the monopoly on fast train development. But this aspect of the HST should not be overplayed, and it would certainly be wrong to give the impression that the HST was designed and developed in total isolation from all the work that had taken place for the APT. Even though the design brief of the HST was for a rapidly developed and overtly 'conventional' train, it could never have been built but for the research work that was undertaken for the APT, and the influence of the APT on the design of the HST is all too apparent.

For many years BR had experimented with diesel trains capable of 200 km/h but 'hunting' vibrations and problems of excessive track wear and damage meant that these designs were not commercially viable. The research into bogie/suspension designs for the APT (outlined in Chapter 4) had their first application in the HST and hence made operational speeds of 200 km/h commercially possible. As such, although the HST followed the evolutionary 'Design—Build—Test—Modify—Introduce into Service' sequence, in the crucial area of the bogie/suspension system, the HST was able to start from a solid basis of scientifically based work undertaken for the APT.

Although the HST project was authorized in order to maintain the commercial impact of fast trains, the development brief still left the design initiative with the engineers of the Chief Mechanical and Electrical Engineer's Department since all that was requested of them was the best 'state of the art' seven-to-eight coach train for BR's non-electrified routes in as short a development period as possible.

Design features of the High Speed train

The design of the 200 km/h HST is superficially similar (save for it not tilting) to the gas turbine APT-E (see Figure 18). This indicates the design influence of the working relationship that had developed at the RTC between BR Research and the CM & EE Department. Like the APT-E, the HST consisted of double streamlined power cars located at either end of a set of passenger coaches, thus making up a 'fixed formation' train. As with the APT, it was necessary to use double power cars as a single diesel engine would have been too bulky and heavy. Compared to the APT, the HST power car output was relatively low (1.7 MW per car as opposed to 3 MW on the APT-P), but it was the use of diesel engines, with their higher weight-to-power ratio than electric traction, that resulted in the need for two power cars for quite a small train (375 to 447 seats).

Gas turbines were briefly considered to power the HST,

Figure 18. The prototype High Speed Train
Source: British Rail

but in practice a high performance version of an existing diesel engine was used. Gas turbine power was not necessary to achieve 200 km/h, and fuel consumption would have been much higher than that of a diesel engine. Also, the use of diesel fitted in with the rapid development brief of the HST, which did not allow much time for prototype development and thus positively discouraged all but the most essential innovations. The HST's designers did not want to push too many design elements too far too rapidly.

The Valenta engine used was a development of the existing Paxman Ventura, which was already in service with BR. The engines were designed to produced 1,865 kW (2,500 hp), the increased power being achieved by a new design of turbocharger and fuel injection equipment. A number of other improvements in the build of the engine were needed to cope with the increased power, loadings and temperatures. As with the APT, the need for streamlining and weight saving was important in order that the running costs of the train should be commercially acceptable.

The approach used on the HST displays a mix of APT-type methods and innovations in the more evolutionary tradition. An example of the latter is in the design of the 'Mk 3' coaches for the HST. These were conventional, non-articulated steel coaches which had been under development anyway for existing trains and were modified for application in the HST. Weight saving was mainly achieved by making the coaches longer, so that the seating capacity of each coach was increased. As with the APT, use was made of Glass Reinforced Plastic (GRP) to reduce weight and maintenance costs. Each 32-tonne Mk 3 coach incorporates 1.6 tonnes of GRP, including seats, panelling and partitions. The power cars also include, 1.3 tonnes of GRP each, largely in the streamlined nose-cones of the train. The underside of the Mk 3 coach was designed to be more streamlined than earlier stock, using modular compartments for braking, air conditioning, batteries and the motor alternator. These modules, which can be removed for ease and speed of maintenance, are totally enclosed in a panelled frame, giving a smooth, aerodynamic line to the coach.

As with APT, an important feature of the HST's design was the reduction in unsprung mass. Existing trains had their traction motors mounted on the axles, i.e. on the unsprung part of the train. The HST used bogie-mounted traction motors with transmission down to the driving wheels via a flexible cardan shaft allowing for a 30 mm rise and fall between the axle-box and the traction motor. This greatly reduced the unsprung weight of the HST, which is so important in reducing the track wear and damage of fast train operations. Bogie-mounted traction motors had been used in the design of the Class 87 175 km/h (110 m.p.h.) electric locomotive, the development of which just preceded the HST. As such, it was a relatively straightforward task to adapt this design for the HST. Other features that reduced unsprung weight were the use of lightweight solid wheels and hollow axles. The hollow axles were not as complex as those for the APT (which housed the hydrokinetic brakes). High performance disc brakes were considered adequate for the HST and were able to stop the train from 200 km/h in the same distance as a conventional train using tread brakes at 160 km/h. At 2.24 tonnes per power car axle, the unsprung mass of the HST was higher than for the APT, but with its lower operational speed, track damage was comparable. Overall, production High Speed Trains weigh 389 tonnes, some 18 per cent lighter than existing trains of the same passenger capacity.

Basically, the HST incorporated only one innovation—the bogie/suspension design that made 200 km/h operations commercially viable. All the other features were in the evolutionary development tradition of BR which were necessary to fully exploit the speed potential of the bogie/suspension design. Although some APT technology was used, this tended to be selected only if it did not conflict with the fast development time-scale and evolutionary design philosophy of the train.

The Chief Mechanical and Electrical Engineer later observed that the HST design process followed the 'KISS principle' (Keep It Stupidly Simple): 'Whenever it was suggested that additional control or protection equipment should be introduced, the requirement was carefully examined and rejected unless found to be essential.' [G.S.W. Calder in a discussion of a paper by Sephton, 1974.]

Development and production of the High Speed Train

The development programme of the HST was delayed a year by the same industrial dispute over single-driver operations as affected the APT-E. Trials began in June

1973, and the prototype succeeded in running at up to 230 km/h (143 m.p.h.) and in so doing set a new world record for diesel traction. An initial production run of twenty-seven train formations was ordered within months of the trials starting and the prototype itself entered passenger service from 1975. Technical modifications between the prototype and production trains were relatively minor. Bruce Sephton, who as Traction and Design Engineer in CM & EE was largely responsible for the development of the HST, summarized them as follows:

> Some trouble has been experienced with the engine and turbocharger, but it is hoped that this has now been overcome. The engine sump fractured but the design has now been improved to eliminate fillet welds. [Believed to be the cause of the fracture.] A great deal of effort was devoted to ensuring that the engine cooling system was leak-free, with considerable success. Some trouble has, however, occurred in the form of leakage of the oil tank supplying the Voith transmission; this is being redesigned. On the power car, there has been excessive wear on the disc brake pads and this is under investigation to ascertain if pads of a different material are an improvement. The hydraulic hand-brake on the power car has also been the source of some trouble due to difficulty in keeping the system free from dirt particles in the air.
>
> In the coaches during heavy braking, there have been complaints of smells from the disc brakes being taken in through the air conditioning system and arrangements have therefore been made to close the air inlet ducts when braking above a certain level is applied. [Sephton, 1974, pp. 27–8.]

There were some problems with the uprated engines, particularly the new turbochargers which required a number of modifications, but in his 1974 paper Sephton did not seem to consider the engine problems to be of much concern and talked of uprating them to a total of 3,730 kW. In practice, as is discussed below, there were considerable problems with the engines of the HST, with them actually being downrated for passenger service.

The most substantial design modification was in the train's nose-cone, which was redesigned for the production fleet in glass reinforced plastic to produce a more aerodynamic shape and to incorporate room for a second driver. (The outcome of the dispute over the number of drivers was an agreement for single drivers at up to 160 km/h and two drivers when trains were operated at above this speed.) This agreement posed the HST's industrial design consultant, Kenneth Grange, a tricky problem: how to retain the aerodynamic shape of the train's nose-cone (which had been carefully developed for the prototype using a wind tunnel), while providing for a second driver:

> Our aerodynamics relied on a smooth flow of air to left and right of the front window. Now extra window width was going to be necessary so that both men could keep the original angle of vision. The airflow would consequently be hindered by the sharp meeting of a wider window and the cab body. Alternatively, an obstructive central bar could be inserted to cope with a large window area.
>
> I ran through a variety of options until two pieces of good fortune came to my rescue. First, the glass makers announced that they could increase the maximum size of glass, allowing us to abandon the central window bar. This produced poor airflow over the sharper corners, so a principal problem remained unsolved.
>
> The second stroke of luck occurred when I asked—with great trepidation (because you don't put childish questions to Chief Railway Engineers)—'Tell me once again why we have to have buffers?' To the

> man's great credit he admitted that in this train there was much less need than normal. So that was my breakthrough! The buffers were dispensed with, and the airflow was diverted from around the sides to over the top. The result was a more stylish appearance, sounder aerodynamics and an effective modern image for Britain's railways. [Grange, 1983, p. 47.]

If you compare the prototype HST in Figure 18 with the production version in Figure 19 you can see how the nose-cone design changed (including the removal of the buffers!).

Owing to the pressure of time, some modifications were directly incorporated into the production models without being evaluated on the prototype. Given that the changes were few and relatively minor, the risks were not seen as great. Use was also made of the research facilities built for the APT in order to reduce development time. This

Figure 19. An HST 'InterCity 125'
Source: British Rail

included static loading tests on the power car bogie frame and the coach mainframe and vibrational tests on the power car to establish the natural frequency of components. The latter was to avoid problems of vibrations in the train coinciding with the natural vibration frequency of components, so causing them to be damaged or become faulty.

Production High Speed Trains were manufactured by British Rail Engineering Ltd (BREL) and entered general passenger service on the London–South Wales route on 4 October 1976, marketed under the name 'Inter-City 125' to emphasize their top speed of 125 m.p.h. East Coast main line service began in May 1978, reducing the best London to Edinburgh time from 5h 27 min to 4h 35 min. Production continued until 1982 with a total of 198 power cars being built. Ninety-four trains are now in use, allowing for ten power cars to be in reserve or undergoing maintenance.

Because of the HSTs' inability to take curves any faster than slower trains, they were initially introduced, as mentioned above, on routes with relatively straight track where curve speed constraints would not severely limit a non-tilting train (e.g. London to South Wales and the East Coast main line). On such routes the HST can achieve *average* journey times of over 160 m.p.h., and when they were introduced the only train to beat the HST for speed was Japan's Shinkansen. Since then the HST has been nudged into third place with France's 270 km/h TGV taking over top position. The HST has since been introduced on other routes where, although its 200 km.p.h. capability is not fully exploited, faster journey times have been achieved because the train is permitted to exceed normal track speeds. This is because of the HST's high power, good braking performance and the fact that its low unsprung weight and weight distribution create less track wear than existing trains. As with the TGV, the benefit is spread over a large network of lines, but at generally lower speeds and with a stronger emphasis on the technology of the train allowing it to run faster on existing track rather than relying on major track improvements.

Passenger response to HST services

In the first four years of HST service on the London to Bristol and South Wales route (1976–1980), the number of passengers increased by 40 per cent. In the first year alone of HST services to Leeds, passenger growth was 13 percent. A HST monitoring exercise was established in order to precisely identify the extent to which this increase in passenger traffic was attributable to the improvements in journey time brought about by HST operations. The earlier models, based on the experience of the electrification of the West Coast main line in 1967/68, had tended to assume that the entire growth in passenger traffic had been due to improvements in speed and frequency. The forecast of passenger response to the HST had been based on such models. Shilton (1982), in assessing the HST monitoring exercise, asked whether such models were attributing to speed the effects of other factors:

> Rail carryings, particularly on Inter-City routes, are susceptible to movements in the national economy, and of course volume will be affected by periodic fares increases. . . . For the HST monitoring exercise, the first step was to make reliable estimates of the growth in traffic on each flow [i.e. route] following HST introduction. The method used to do this was to search for flows—'control flows'—which had behaved similarly to a particular HST flow in the past and had not experienced a service change, nor were about to do so. It could then be assumed that any difference in future behaviour could be attributed to HST. [Shilton, 1982, p. 791.]

The process was then taken one stage further, examining the relative influence of all the changes that the HST service incorporated: 'the quality of the coaches, the image of the HST, the extra advertising it received and, perhaps, an increased awareness by the public of the service on offer' (Shilton, 1982, p. 721). Although it is difficult to quantify such factors, they were found to account for roughly half of the increased patronage associated with the introduction of HST services.

A similar effect was noted with the TGV. SNCF had assumed that, to be competitive, the TGV would have to run over the new line for at least 60 per cent of the time. Yet the traffic to Geneva and Marseilles, involving a majority of old track, was far heavier than was expected using simple speed elasticity modelling. Only now (1986) are the French undertaking studies to identify what factors other than speed and frequency affect TGV traffic.

Overall, Shilton concluded that the growth in passenger traffic was accurately predicted, but probably for the wrong reason:

> Whilst the growths which have been observed when the HST was introduced are of the same order as was expected from earlier time-series work, it now seems that rather less of the overall growth is due to speed and frequency than was at first thought and rather more to the associated marketing effects that the speed improvements make possible.

Industrial design in British Rail

One striking feature of the HST was the use of industrial design to a degree not previously seen on British Rail. Indeed, the High Speed Train won a Design Council Award in 1978. The overall design image of the train has been very strong and has contributed significantly to its success in attracting passengers. British Rail's Design Panel was established in 1956 in response to widespread criticism of the poor aesthetic and amenity standards of the rolling-stock and locomotives introduced under the 1955 Modernization Plan (Haresnape, 1978). It began to make an impact as a result of one of the few positive ideas of Dr Richard Beeching during his period as BR Chairman—that British Rail should have a modern, comprehensive design image through the creation of a distinct house style and colours for virtually everything on the railways, from the livery of trains and the design of seating, down to the lettering style on tickets. The successful marketing of the HST, with it being given a modern and innovative image, owed much to these design aspects. The train had a distinctive livery and the internal design of the coaches very much enhanced the 'special' image British Rail wanted for the train. Much of this was derived from design work carried out for the APT and, somewhat ironically, from 1983 even the APT's distinctive livery has been used for the HST.

Kenneth Grange, the Design Panel's consultant industrial designer on the HST project, has commented on how the design image of this one train has benefited British Rail as a whole:

> Designs like the High Speed Train have a variety of values, one of which is deep in its effects. When a company is in retreat, torn by arguments and ridiculed by the press, the morale of its workforce is soon eroded. This in turn affects attitudes to customers, and the downward spiral accelerates. The High Speed Train—in reality only a small part of the vast efforts made to improve British Rail—has re-awakened some of the old railway pride. Ticket collectors buy and wear HST lapel badges, rebellious station masters paint HST arches, while British Rail posters for anything and everything use the new train as a symbol of progress. [Grange, 1983, p. 47.]

This use of the HST is similar to the objectives set for the

TGV design consultant discussed in the previous chapter. British Rail have capitalized on this, using the train in circumstances where it is the HST's *image* rather than its speed that is the main traffic-generating factor. In particular, some HSTs have been used as 'flagships' for certain routes in order to raise awareness of rail services in general, even though the HST may only run once a day. The 'Cotswold and Malvern Express', introduced in 1984, the once-daily 'Humber–Lincs Executive' and the 'Highland Chieftain' to Inverness are all examples of flagship marketing, exploiting the image and prestige associated with the HST.

The XPT—Australia's High Speed Train

Throughout the 1970s, the state of New South Wales invested heavily in modernizing the railways owned and operated by the State Rail Authority. This increased investment was coupled with a bold decision, in 1976, to reduce fares. Within four years, passenger journeys had increased by 52 per cent with passenger revenue up by around a third. The impetus behind this state-backed rail revival was a combination of energy conservation, the desire to maintain a viable rail network plus the belief that rail subsidies could be reduced using a high investment /low fares policy. Needless to say, this stance has been controversial, but it has worked. Probably the most politically controversial element of this programme was the decision, in 1979, of the New South Wales Public Transport Commission to buy a fleet of four Australian-built High Speed Trains. Called the 'XPT', the basic HST design was considerably modified by its manufacturers, Commonwealth Engineering (Comeng), for service in New South Wales (Figure 20). The XPT is smaller than the HST, each train consisting of five carriages (based on a Comeng trailer design) and two power cars. Accommodation is for

Figure 20. The XPT
Source: NSW State Rail Authority

288 first-class passengers. Unlike the HST, and the XPT design is for a small fleet of luxury fast trains, catering for the top end of the rail market, directly competing with the airlines.

At that time, the maximum rail speed achieved anywhere in Australia was 120 km/h (75 m.p.h.). In New South Wales the quality of the track and severe gradients involved (particularly over the Blue Mountains) ruled out any further improvements using conventional locomotives. The attraction of using the HST design was that, by adapting this 200 km/h train for a smaller train at lower speeds, coupled with a modified bogie design for New South Wales track conditions, the XPT could run at 160 km/h on ordinary track. The impetus for the development of the XPT was therefore very similar to that behind both the APT and HST in Britain. Despite the political backing of the railways, the construction of high-speed lines or even the rebuilding of existing track for fast running was totally out of the question for just four trains. Adapting HST technology to Australian conditions made fast running on ordinary track possible.

The potential offered by licensing the HST design was far from appreciated when the decision to build the XPT was announced. It was viewed by many as an insult to Australia's rail industry:

> With the State elections due to be held in September 1981, XPT became the target for much political in-fighting between State Premier Neville Wran and the opposition Country Party . . . who . . . claimed that XPT was an expensive folly which would never reach 140 km/h (87 m.p.h.). In reply, the previously untested equipment (one power car and two completed trailers, plus a test van), in the course of several sustained trial runs in excess of the design speed of 100 m.p.h., attained a maximum of 113.5 m.p.h., which says much for the fundamental soundness of the Anglo–Australian design. Come the election, Neville Wran was re-elected. [XPT—New South Wales' Political Train, *Modern Railways*, 1983.]

The original fleet of XPTs entered service from May 1982, with a further build being authorized a month earlier in order to increase the size of the trains to six carriages and increase the fleet to six. The commercial success of the XPT was almost immediate and in 1985 this was reinforced by a marketing strategy which included fares reductions. Following this, XPT passenger traffic increased by a third and a further build of carriages was authorized to increase the train size to seven coaches. The State Rail Authory has now ordered a further fourteen sets. With the recent authorization of the final link in the Transcontinental railway linking Alice Springs with Darwin in the Northern Territory, a second generation XPT is being openly canvassed for this route.

The role of fast trains in Australia is clearly smaller than in Britain. With such a restricted market, the design constraints are very severe since virtually no specific infrastructure investment is viable. The XPT *had* to be able to run on track which had no special adaptations for it. The Japanese, French and American approaches were all ruled out. Conceived in the atmosphere of Britain's economic and political constraints on rail investment, the HST provided an almost ideal design from which the XPT was developed.

Technical and operational limitations of the High Speed Train

The High Speed Train is certainly a success story, but it is not an unqualified success. Problems have been experienced at two basic levels: (a) technical; (b) operational and design inflexibility. Technical problems have focused on the maintenance and reliability of the Valenta engine, which, although adequate for commercial services, is poorer than expected and in consequence the trains are more expensive to run than was anticipated. There were, and still are, a variety of problems. The lightweight aluminium gearboxes suffered from cracking, so steel replacements were fitted to the entire fleet of HSTs (the XPT had steel gearboxes from the beginning and so has not suffered from this problem). There has been a persistent problem with internal coolant leaks, plus a variety of miscellaneous mechanical failures. One early response was to downrate the engines from 1,865 kW to 1,680 kW in an attempt to reduce maintenance problems, but many still persist. The problem with the power car disc brake pads cracking was never solved. They just have to be replaced far more frequently than was envisaged. The fitting of rheostatic brakes to four HST power cars was undertaken as a pilot scheme in 1985.

Table 2 Estimated operational cost (i.e. excluding labour) for the High Speed Train and Class 87-hauled electric train

	8-coach HST at 200 km/h (450 seats)			10-coach Class 87-hauled train at 175 km/h (590 seats)		
	Per year	Per mile	Per seat mile	Per year	Per mile	Per seat mile
	£	£	p	£	£	p
Capital	283,000	1.32	0.29	253,000	1.27	0.22
Maintenance	488,000	2.28	0.51	311,000	1.55	0.26
Fuel	293,000	1.37	0.30	208,000	1.04	0.18
Total	1,064,000	4.97	1.10	772,000	3.86	0.66

Source: British Rail. Prices are as at October 1984.

The basic response to the technical problems of the HST has been to increase levels of maintenance to get reliability up to a reasonable level. Each HST now costs about half a million pounds per annum to maintain—twice as high per seat mile than other trains (Table 2).

In 1983, following a series of technical problems with the HST, BR reduced its HST fleet in order to have more power cars available in reserve. The availability of HST sets has subsequently improved and service problems are less frequent than they were. The idea of totally replacing the engines is being seriously considered, such have been the problems with the Valenta. In 1986, four power cars were re-engined with an alternative diesel, the Mirless MB190, but the greater weight of this engine increases axle loading which would probably require a reduction in top speed in order to keep track forces down to an acceptable level. For this reason, any re-engining seems more likely to be associated with the eventual transfer of HSTs to slower, secondary routes. Meanwhile, British Rail has obtained £7 m in compensation from the Valenta's manufacturers in recognition of the technical problems with this engine design.

To some extent, this problem is not purely technical, but relates also to the very intense use that is made of the HST sets in order to meet financial investment targets. It is interesting to note that there have been no substantial technical problems with the XPT in Australia, which is not utilized at anything like the level the HST is in Britain and also operates at lower speeds. An overall technical assessment of the HST is that it is a successful train, but this success has been bought at a higher price than was expected.

Operational flexibility

The operational problems, or rather limitations, of the HST relate to the basic design concept which it shares with the APT—that of a fixed formation passenger train. The power cars cannot be used separately from their coaches like conventional locomotives and hence cannot operate other services such as parcels, freight or sleeper trains. Given that these are predominantly night-time uses, they actually require a separate fleet of locomotives. High utilization is an essential factor in railway economics, and so long as a fixed formation passenger train is well used, this inflexibility does not matter. Indeed, fixed formation trains have a lot of advantages: they can be driven from either end, maintenance can be standardized and they have a very attractive marketing image, the latter being particularly notable with the HST. Hence, not only BR opted for fixed formation passenger trains from the early 1970s, but all other fast train developments elsewhere in the world have been of this type.

The main problem with fixed formation trains arises when utilization drops below around 400,000 km (250,000 miles) a year. In Britain, with tight financial restrictions on rail investment, this constraint is even stronger. The HST production run was slightly curtailed as further investment in such dedicated train sets was not considered viable. With the onset of the economic recession in the late 1970s and resultant changes in BR's passenger markets, this fixed formation concept of a train specially designed for one purpose rather than able to be used for several purposes within the railway became less attractive. This was to prove of particular relevance in determining the ultimate fate of the APT project.

More immediately, for passenger traffic operations, the fixed format of the HST was to provide a serious constraint on its use. The number of coaches could not be changed easily, the trains could not be split to serve two branches from one main route, the ratio of first to second class accommodation was fixed, even though the relative demand may vary if an HST train set operates on more than one route in a day. Extra coaches were slotted into the train sets, and, when some coaches were refurbished, a higher density layout was adopted to increase seating capacity to seventy-six. Even so, the design of the HST was not flexible enough to cope with the seasonal, weekly or daily fluctuations in demand experienced on some routes. There have been examples of HST services being replaced by slower conventional trains because of their lack of capacity or inflexibility. For example, on the routes to Cornwall HSTs have been replaced by larger and slower conventional trains which are more appropriate for predominantly holiday traffic. The HSTs have been moved to other services more appropriate to their capacity and performance.

One embarrassing example of the inflexibility of the fixed format HST to adapt to changing market needs was in the provision of baggage space. In the HST there is no parcels van as such—parcels and baggage being accommodated in a small compartment to the rear of each power car. Soon after the introduction of the HST, British Rail Launched its 'Red Star' parcels service and also made the carriage of bicycles free in order to promote leisure use of the railways. The tiny baggage compartments of the HST could not cope. Parcels vans could not be added on to the trains and so bicycles were banned from most HSTs in order to provide room for the more lucrative parcels traffic (this has since been replaced by a three bikes per train reservation scheme). This problem was not unique to the HST, for most new local trains designed at that time were equally incapable of coping both with bikes and an increase in parcels.

The inability of the HST to adapt to some BR marketing innovations and other changes has limited the commercial impact of these innovations and the use of the HST itself. The extent to which this was due to its design concept being derived in relative isolation from other decision-making in BR, or was simply because of changing market conditions leading to unforeseeable requirements, is a matter of debate. BR did establish a *High Speed Steering Group* with representatives from all relevant operational sides of BR to meet with both the APT project team and those building the HST. The managerial structure therefore did exist to feed in operational and marketing requirements to the train's design and, overall, is considered to have worked successfully. The main problem with both the HST and the APT was probably that they were basically 'lean' designs, developed for a particular concept of the rail passenger market. Rothwell, Schott & Gardiner (1985, pp. 27–8) argue that flexible and adaptable 'robust' designs, though not ideal for any one task, will succeed better than highly refined 'lean' designs, which are built so carefully around one market that they cannot cope if there are mistakes in identifying that market or if the market simply changes.

With the onset of the 1980s recession, government cuts, coach competition, etc., the rail market has changed a lot and a train designed for one concept of the future is finding it difficult to cope with another. It cannot be claimed that the HST design was as lean as, say, that of

Concorde. The HST is now used for a number of purposes other than that for which it was originally conceived and, particularly given its very short development period, any commercial shortcomings are mere qualifications of an overall success story. Yet the areas in which problems were experienced were to be magnified in respect to the HST's 'high tech' brother, the APT.

7 Managerial Haemorrhage

We left the Advanced Passenger Train at the point where, having proved the concept of a scientifically developed high-performance train to be sound, the project was moving from an experimental into a design and development stage. This chapter considers the technical and, in particular, the organizational and managerial factors that resulted in the prototype APT-Ps failing to enter full passenger service in 1981. This crisis forced a reorganization of the APT project which eventually produced a usable passenger train. But by then market conditions had changed and the APT was no longer appropriate. The prototypes became test-beds for technology to be incorporated into a new design of train—which is the subject of Chapter 8.

Organizing the development of the prototype APT-P

With the BR Board's approval of an electric APT in 1971, a small group from BR Research had been seconded to work with the CM & EE Department engineers to review the technology and recommend how to take the project forward. The Joint Review Team's proposals for the APT-P were submitted to the BR Board in April 1973, but the authority to construct prototypes was to take seventeen months to come through (nearly as long as it took to develop and build the prototype HST!). This delay was largely connected with a wrangle between the BR Board and the Government over the financing and scale of the APT project. The Department of the Environment (into which the Ministry of Transport had been merged in 1970)

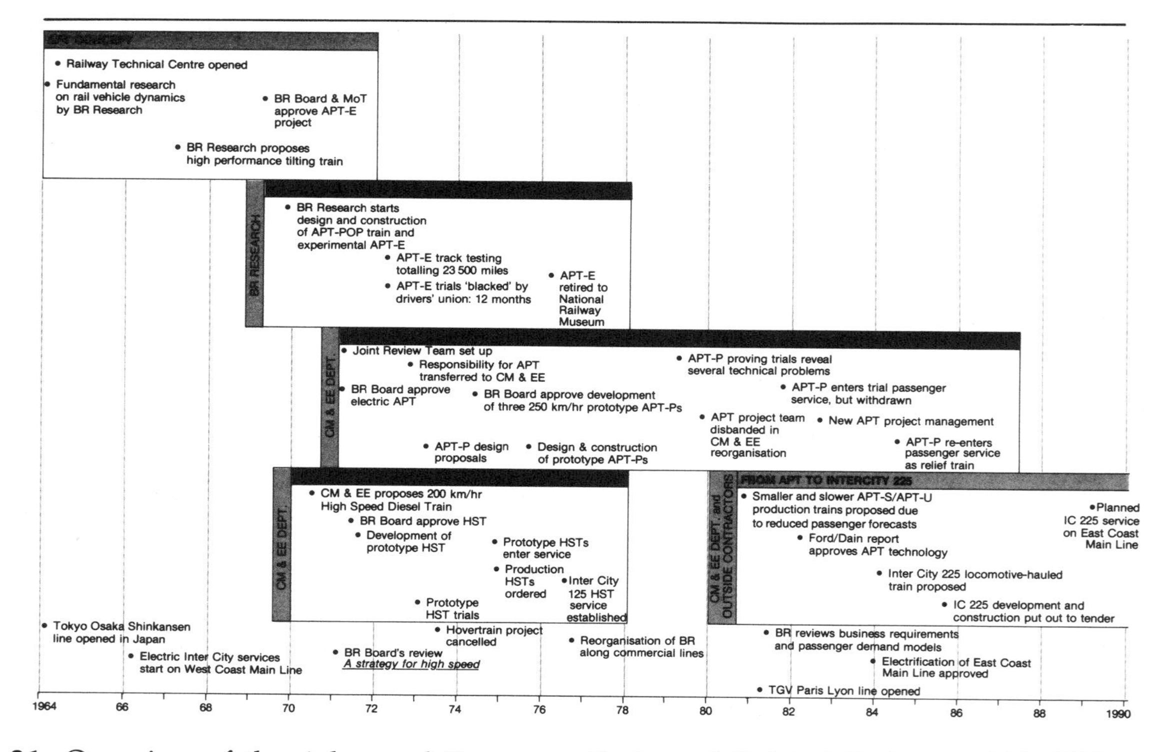

Figure 21. Overview of the Advanced Passenger Train and Related Projects, 1964–1990
Source: Potter & Roy, 1986, p. 41

was by now very enthusiastic about the APT project. The APT-E had aroused a lot of international interest, raising the prospect of substantial exports. For this reason the Department of the Environment proposed that ten prototypes be built and were willing to provide 80 per cent of the money needed.

In the wake of the 1971 *Strategy for High Speed* and the rapid and successful progress on the HST, the BR Board were not prepared for such a substantial commitment. They had four trains, at most, in mind. In the end the Department of the Environment's offer was lost when, financed by internal resources alone, authorization for only three APT–P trains was given in September 1974. This scaling-down of the APT project may well have discouraged commitment to the APT by other BR departments and private industry, a problem that became very real.

At that time it was envisaged that eighty production APTs would be in service from 1980, with more to follow as routes were electrified. It was planned to have the prototypes operating passenger services on the London–Glasgow route from 1977, reducing journey time from five to three and a half hours.

The project was now the responsibility of the Chief Mechanical and Electrical Engineer's Traction and Rolling Stock Design Department. The core of the APT Project Team, about thirty people, were transferred from BR Research into CM & EE in April 1973. The group eventually grew to about seventy to eighty strong, plus forty outside contractors. At its height, the APT team within CM & EE constituted 10 per cent of the Department's total staff at the Railway Technical Centre.

However, the concept of a project team was totally alien to the way in which the CM & EE Department was organized. This was structured on a 'functional' basis, with the Traction and Rolling Stock Design Department having separate offices for locomotives and carriage/

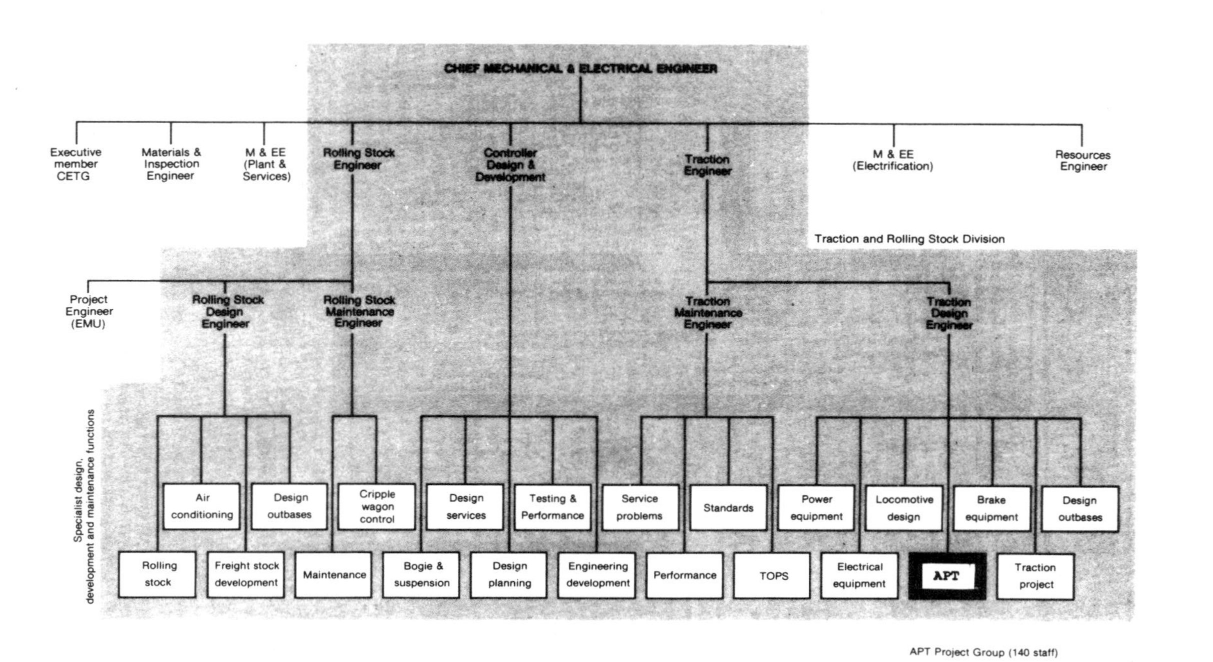

Figure 22. The CM & EE Department as organized on ‘functional’ lines in 1977
Note the way in which the APT was the only separate group for a specific project
Source: Potter & Roy, 1986, p. 42

wagon design containing specialist engineers for electrical equipment, power equipment, brakes, air conditioning, etc. (see Figure 22). Work was allocated by the Chief Mechanical and Electrical Engineer via his senior staff to the Traction and Rolling Stock Design Engineers, and it was their responsibility to co-ordinate all the specialist jobs in their section. A train design and development job would normally be just one of many booked in and seen through the various parts of the Traction and Rolling Stock division by the senior engineers under the Chief Engineer. This functional structure was entirely appropriate for the evolutionary engineering philosophy of the railways. There was no concept of allocating people exclusively to one job, or to grouping people into project teams. But the APT was organized on just such a project team basis. The APT-P Project Team contained specialists with skills in all the design areas necessary for the train's development. This caused problems: these specialists duplicated the engineers already in CM & EE and in many cases this was resented. Also, the imposition of a project team structure served to encourage the isolation of the APT Project Team from its new department. Finally, scepticism about the whole APT concept still persisted among several CM & EE Engineers, especially given their success with the HST, the first production order for which had just been placed.

The problems of transforming the APT from a research project into a practical development had not been overlooked. The decision to transfer the APT Project Team to CM & EE and to retain the project team structure was viewed as the best way forward to implement a radical concept within an organization geared to evolutionary technological progress. The dangers and problems of imposing a project team on CM & EE were recognized, but this was considered the most practical way forward.

Such problems are not, of course, unique to British Rail or the APT. They face any large organization attempting major innovative projects. Consequently, the structural features that affect an organization's ability to innovate have received considerable attention from management scientists. Oakley (1984), for example, discusses why departments which have a functional structure, to efficiently carry out routine tasks, tend to be 'mechanistic' and unsuited to activities like innovative design, which require a more flexible 'organic' structure. The greatest barrier faced by the APT was not technical. It was managerial. Specifically, the problems were:

- It was a radical innovation in an industry which had long been organized with evolutionary developments in mind.
- Almost following on from the above, it was a project which, because it challenged established orthodoxy, inevitably became the focus of a lot of internal rivalries, particularly when it moved into the department where conventional trains were designed.
- Because it was a radical development it required a different form of management (in this case, the project team), while the usual evolutionary design work continued in the functional departments. The imposition of one type of management on top of another caused overlaps and uncertainty in the responsibilities of staff.

Given the nature of the APT project, it is hard to see whether, at that time, any alternative organization for its development would have been preferable. One method considered, which certainly would have been favoured in today's political climate, was to set up a company separate from CM & EE and British Rail Engineering Ltd (BREL) to develop the APT. But in 1974 such a move was not acceptable to the BR Board, the rail unions or the Government. One major organizational constraint was that the APT would have to be built using the accepted development and manufacturing structures of the rail industry.

At the opposite organizational extreme, merely passing over the project to CM & EE without the project team would have produced immense problems in implementing a radical innovation in an organizational structure totally geared to evolutionary developments. The whole success of the APT up to this time had been due to the commitment of BR Research and the use of a project team so that the implications of any one design change could be followed through to other areas. Given that a 'functional' organization is unsuited to radical innovations, the APT project would probably have got bogged down or been killed off by its opponents in CM & EE.

In practice, therefore, a mid-way path was chosen. The project team moved to CM & EE to lead a somewhat uneasy existence. The interface between scientific research and practical rail development was imperfect, but was viewed as the best that could be managed at the time.

Trials . . . and tribulations of the APT-P

The construction of the APT-P prototypes was undertaken under the direction of the project team by some thirty subcontracting companies. Bodies and bogies were made by British Rail Engineering Ltd (BREL) and traction equipment by the Swedish firm ASEA. It was in the BREL workshops at Derby that the components and trains were assembled. The delay in authorizing the construction of the prototypes combined with delays in their actual manufacture by BREL resulted in the programme slipping, although of course design work had been under way since 1973. The first power car was delivered in June 1977 and underwent track testing at up to 200 km/h later that year. The trailer cars were delivered in June 1978 and these, like the power car, initially underwent track testing on their own in July 1978. In late 1978 another industrial dispute with development staff and drivers delayed the development programme, so it was not until February 1979 that the first complete prototype train was ready for proving trials, when it was delivered to the Shields Depot, Glasgow. Track proving trials finally began in May 1979 and in December 1979 this train set another British speed record at 260 km/h (161 m.p.h.), although signalling (as will be recalled from Chapter 4) generally limited the train to 200 km/h. The second prototype was commissioned in late 1979 and the third in 1980. Although these trains were able to make a series of very fast runs, there were considerable problems with tilt failures, binding tread brakes and the gearboxes. In order to test the trains as a whole system it was necessary to use them in passenger service. But to do this, the trains had to be technically reliable.

The lack of passenger trials became a source of embarrassment to BR and increasing pressure began to be applied to the APT project team to get the trains into service. The Department of Transport (which in 1974 was separated from the Department of the Environment) would only authorize a production run of APTs once the prototypes had proved themselves in actual passenger service.

In 1980 BR went out on a limb and featured the APT in the published passenger timetable on the West Coast route. This involved one train making three return trips a week at 200 km/h between Glasgow and London in a scheduled 4h 15 min; but the service was postponed while work continued on sorting out the reliability problems on the trains. Finally, on 7 December 1981 a single APT-P entered passenger service. This first run, from Glasgow to London, although officially limited to 200 km/h, in fact reached 222 km/h (138 m.p.h.), setting yet another British speed record for a passenger-carrying train. The journey was completed in the scheduled 4h 15 min, but the glory was short-lived. On the return trip the driver took a bend too fast, amounting to 16° of cant deficiency. The APT-P is fitted with a safety device, such that if there is a wide tilt angle for any length of time it is assumed that the carriage has stuck in the tilting position and the self-righting and locking mechanism is activated. Taking this bend too fast activated this safety mechanism, causing all the coaches to revert to the upright position. On its own, this problem was by no means a major one, but then a fault developed on the main traction circuit which blew the main circuit breaker, shutting down the tilt mechanism altogether. The train had to slow down for bends and so was late in at Glasgow.

Two more attempts were made to run the train, but added to the APT's technical problems were the worst weather conditions for thirty years. For the second run, on 9 December, the train set off from Glasgow in a blizzard with temperatures at between −15° and −19°C. The temperature dropped even further when, climbing Beat-tock summit in southern Scotland, the train was brought to a halt. Small amounts of water in the tread brake air hoses between the carriages had frozen and blocked the pipes. This tripped off a safety mechanism which detects when air pressure falls below an acceptable level and automatically brakes the train. Basically, like many trains on the BR network that morning, the APT got frozen up. But no other train attracted as much publicity in becoming iced-up as did the APT.

A third run was then attempted and the APT got as far as Crewe successfully. This in itself was quite an achieve-ment, since by this time heavy snow was drifting across the track. But south of Crewe the snow was even worse. The track was blocked and no trains at all were moving. The APT therefore simply set off north again, picking up snowbound and stranded passengers as it went. One mile from Preston the train came to a halt. An electrical short on a minor circuit had (through incorrect wiring) causing a short in the traction and train control circuits. The APT had again to be towed. The fault was corrected at Preston in under an hour, after which the APT continued up to Glasgow.

The publicity associated with these attempts to operate the APT very much embarrassed BR. Unwilling to see such a prestigious train operate unreliably, the APT was withdrawn from passenger service. It was due to return in May 1982, but never did. In September 1982, British Rail finally admitted that the APT-Ps would not see regular passenger service, but that they might be used as relief trains or in peak summer periods. Trials of the APT continued, using BR employees as 'passengers' and in August 1984, fifteen years after the project began, one APT eventually returned to passenger service as a relief train on the London to Glasgow route.

Why did the APT fail?

The delays and eventual failure of the APT-P to enter into passenger service is a saga that is familiar to many people. But what was really behind the APT's problems? Clearly there were technical failures, but are these alone an adequate explanation, or were the technical failures on the APT merely symptoms of a much deeper problem of managing a complex, innovative project within a large organization?

Technical problems

Clearly there were technical problems, and perhaps the best way to tackle this question is to begin by looking at these. The main ones were:

1. tilting mechanism jamming or failing;
2. tread brakes binding and over-heating;

3. hydrokinetic brake problems causing axle failure;
4. poor ride of articulated bogies;
5. gearbox failures

The problems with the tilting mechanism were experienced early in the development of the APT-P and persisted throughout the project. One early problem was that the tilt response was found to be rather late and jerky. The sensor activating the tilt was therefore relocated on the preceding coach and this solved the problem. The coach 'anticipated' curves and so was tilting as it entered into them. However, this modification to the prototypes was bought at the price of reduced reliability. The tilt mechanism on the APT-P had been designed as a duplex system, with two parallel sets of machinery and sensors at the ends of each coach. Should one fail then the second would continue to operate. But the placing of the sensors on the preceding coach required a connection between coaches which had not been allowed for in the design. All the wires between coaches were in use, but eventually a connection was sorted out. However, there was not the capacity to provide for the full duplex system involving two accelerometers and two parallel paths for the signals. Hence a single link had to suffice between the accelerometer on one coach and the two tilting mechanisms on the following coach. This was non-duplicated, with the result that a single fault could knock the tilt mechanism out of action. On production trains it was planned to reintroduce full duplication, but for the APT-Ps the reduced reliability of a non-duplicated design had to be accepted.

The binding of the tread brakes proved to be a remarkably persistent problem. The dragging of tread brakes is not unusual; it occurs on conventional trains without causing any real difficulties. What turned this from a minor irritation into a major problem was the combination of a design fault in the tread brakes with the small, lightweight wheels of the APT.

The design fault was simply that the brakes had too small a clearance with the wheels. Tread brakes are simply pads that clamp on to the tread of the wheel to slow the train down by friction. On some wheel-sets the clearance between the braking pads and the wheels was adequate, but it only required a very small error in manufacturing or assembly to result in the brakes constantly dragging. The friction of dragging brakes would cause the small wheels of the APT to heat up very rapidly. The rim of a wheel would become hot and would try to expand. This could pull the hub of the wheel out of fit and loosen the wheel on an axle. The problem was eventually overcome by fitting a device to monitor and identify dragging brakes, plus a restraining collar between the wheel and the axle so that if the wheel did become loose through dragging brakes it would not shift. Any badly dragging brake would then simply burn out without causing other problems. For the APT-P prototypes the dragging brakes remained a problem that the engineers realized they would have to live with. For any production trains it was accepted that a less sensitive design would be adopted, and even disc brakes were under consideration.

The worst technical problem encountered by the APT-P related to the hydrokinetic brakes and axle. A bearing failure in one hydrokinetic brake had led to an axle nearly breaking. As with the tread brakes, this was an example of one failure setting off a series of major unanticipated effects. As the bearing collapsed, its case became loose and rotated against the steel casing of the brake. This friction heated up the central part of the axle and the aluminium flange of the brake. This flange expanded beyond its yielding point so that when the axle cooled the bolt holding the brake and axle was loose. The replacement of the aluminium flange by one of steel cured the problem, but this took a considerable amount of time. In the meantime, with such a potentially dangerous situation arising out of a simple bearing failure, trials of the APT-P trains were virtually halted.

The gearbox problems were basically a lubrication weakness. The design relied on forced lubrication, the details of which are unnecessary for this case study. However, if oil became contaminated, lubrication became ineffective, causing bearings to fail. There were also problems of oil leakage through seals. The failures were fairly frequent and as such held up the general development of the train. The problems were solved by both improving maintenance to reduce contamination and by fitting oil coolers.

Manufacturing problems

But the most disturbing problems on the APT-P trains did not relate to design and technical problems but to the quality of manufacture and assembly. In April 1980 one APT-P, while on a demonstration run at Oxenholme on the West Coast main line, suffered a derailment while travelling at 200 km/h. The train stopped safely and the cause of the derailment was found to be a collapsed wheel-set. A ring of bolts securing the conical axle to the hydrokinetic brake housing had not been tightened. This led them to fracture and the wheel-set had simply come apart. The breakdown of the APT during its third run in public service in 1981 was also caused by poor quality control—a rubber grommet had not been fitted over a wire, which in consequence had chaffed through, causing a short which then shorted out the whole train control because the latter itself had been incorrectly wired up.

In considering problems of manufacturing, a number of alternative explanations need to be considered. These are:

1. Were the design standards expected of the APT sub-contractors and the BREL workshops unrealistic?
2. Was the quality of workmanship on the APT poor? If so why?
3. Even if one (or both) of the above are true, was the APT project organized in such a way as to be able to cope?

In the case of the dragging tread brakes the design problem was simply that the specification did not relate to the standards of construction and assembly of the rail industry. Equally, the design of the gearboxes required maintenance standards that were higher than were available on British Rail.

This raises the criticism that the design of the APT was unrealistic given the manufacturing methods and maintenance facilities of the rail industry. This is an argument that I have certainly come across, which tends to be advocated by those who point to the ex-aerospace engineers on the APT project team. It is not so much a criticism of the rail industry as of a design process that is not fully in touch with the world of which it is a part. Oakley (1984) emphasizes this aspect:

> Many designers have practical experience of production and fully understand the limitations and capabilities that they must work within. Unfortunately, there are also many who do not have this experience and, quite simply, do not appreciate the systems that they are supposed to be designing for. [Oakley, 1984, Chap. 6.]

But, in the case of the APT, although there were some instances of such design faults, these seem more like isolated lapses than a general design malaise. Overall, the APT design did not require any more special attention in manufacture than any other new train project.

A separate, but related, explanation concerns the workshop 'culture' of the railway industry. This suggests that the design standards demanded by the APT were not too high, but required a different style of production than was usual in the rail industry. In older, 'medium' technology industries such as rail (as opposed to 'high' technology industries such as aerospace) there has always been a tradition that the shop-floor can modify and amend design details. Production drawings are very much open to the creative interpretation of the people engaged making a particular component. The experience of such people on the shop-floor is virtually a final check upon the designer. This tradition of 'craft building' did lead to problems, for a number of design details on the APT were changed without reference to the design office because the shop-floor had always done things in a certain way. In some cases it was just a matter of not believing that the lightweight APT designs would work because they were so different from the familiar diesel (and steam) vehicles people were used to.

Although this problem of 'culture' was very real, it only accounted for a minority of the problems encountered on the APT. This therefore brings us on to considering the other explanations: were the standards of construction on the APT particularly poor? If so why? But in actual fact this explanation is closely related to the third: the management and organization of the APT project. The two are therefore considered together.

Organizational structure and project management

The individual technical problems experienced by the APT reflected a basic problem of innovation management and organization. Some of the technical problems were certainly very difficult to tackle and there were some differences in engineering standards and approach between the APT project team and component manufacturers and train assemblers. But there was no reason why these should have produced a total barrier to innovation. It was in the management of the APT project that the true seeds of failure lay.

As mentioned above, the management structure adopted to develop the APT-P prototypes was to move the APT project team from BR Research into the functionally organized CM & EE Department. It was recognized that this would involve organizational problems (not unrelated to the internal rivalries generated by the APT project), but given such a situation this approach was viewed as the best way forward.

In practice the APT project team found themselves in a very difficult position. The Chief Mechanical and Electrical Engineer was very supportive of the APT project, but within the Traction and Rolling Stock Design Department there was considerable antagonism. A number of key engineers viewed the whole approach of the APT as a threat to their professional reputation, their status and method of working. The APT was not only viewed as a frivolous high-tech irrelevance, but as something that was a distraction to valuable staff who could be better occupied in developing and building 'practical' trains. So while the senior management of CM & EE were sympathetic to the APT, at a day-to-day working level there was little support for the APT project. APT work got shunted down to the lowest level of priority. It would take days or weeks for even minor jobs to be done.

The practical effects of this antagonism filtered through to other parts of the railway. The links between CM & EE and British Rail Engineering Ltd were, understandably. very strong. BREL at this time built most of BR's coaching stock and many locomotives, including the HST. As in CM & EE, the senior management of BREL were supportive of the APT, but enormous problems occurred if an individual workshop manager viewed APT-P work as a low priority, 'experimental' job. Such an attitude not only held up the construction, development and modifications to the APT-Ps, as work on 'real' trains were given priority, but it had an enormous effect upon quality control. The difference in quality of work between workshops where the manager was enthusiastic about the APT and where the manager was not, was considerable. Many of the problems of reliability with the APT-Ps were a direct legacy of such antagonisms and attitudes leading to poor quality control.

Once running trials began, problems of lack of commitment to the APT in other branches of BR became apparent. The commissioning team and the APT-Ps were based at an operational depot, working with staff maintaining and servicing inter-city trains. They also had to fit their exacting 200 km/h test schedules into the daily operating pattern of services along the West Coast main line. The image of the APT as a 'distracting high-tech irrelevance' did not help the commissioning team to undertake their work.

Development, testing and modifications to the APT-Ps thus became a very protracted process. It cannot be denied that there were serious technical and design problems, but the main problem was a divided attitude within the whole railway as to the credibility of the APT project. In many key areas this affected the quality of APT work and very much hindered the work of the project and commissioning teams. The APT project team (or Project Group, as it was called) consisted of extremely capable engineers, but no team can possibly contain all the skills and experience needed to see through the development of as major an innovation as the APT. As in all branches of industry (and elsewhere), the team had to rely on colleagues and bosses being aware of how they were getting on and stepping in to assist them in areas of expertise and experience where this was necessary. Designing a high-performance train is one thing, but having the ability to win the support of a large number of departments and key individuals in order to smooth the development and testing process is another.

The APT Project Group lacked the guidance and general support of the CM & EE Department and the accumulated railway experience that such guidance would have embodied. Hence, particularly when other parts of the BR organization were involved, the APT team learnt only from their own mistakes. They had very little access to the practical experience of others. The use of a working depot as a base for commissioning the APT-P trains is a case in point. A working maintenance depot was not the best place to undertake development work, particularly as the prototypes faced considerable technical problems with the brakes and tilt. The commissioning team lacked experience in running trains on an operational track and the staff of Shields Depot were by no means used to undertaking work on the scale that was necessary to debug the APT prototypes. People with a better working knowledge of a maintenance depot and with experience of operative service trains would have advised against such an arrangement. A development project team would not be expected to possess such experience, but would rely upon others within their department for advice. The organizational and personal isolation of the APT team within CM & EE resulted in a lack of such support. Mistakes occurred which should never have happened.

Fragmentation and reorganization

Given this situation, the progress that was achieved on the development, construction and testing of the APT-Ps was remarkable. The judgement that transferring the APT Project Group to CM & EE was an acceptable compromise was vindicated. Progress was slower than planned, there were some frustrations and difficult times, but overall the project was progressing acceptably. Given the organizational constraints within BR, the APT was doing as well as could have reasonably been expected. But three years after the start of development work, just as the APT Project Group were beginning to win the enthusiasm of others within CM & EE and to open trials with prototypes, a new managerial factor came into play that virtually brought any real progress on the APT to a halt. This was simply the total reorganization of the Mechanical and Electrical Engineering Department which, beginning in 1976, took four years to complete.

The objective of this reorganization was to produce a department that related much more closely to the commercial requirements of the railway than did the old structure of functional categories. Throughout the late 1970s and early 1980s the whole of British Rail went through a similar process, producing a management structure geared to rail's main markets: inter-city, provincial services, London suburban services and freight. It is a strongly commercial structure and one which ensures that the working of all departments within the railway relates to the business side of the industry. Commenting on the change in managerial approach that this new structure represents, The *Observer* noted: 'In essence Reid [BR's Chairman] has turned BR from being an organization dominated by the engineers into a customer-orientated market-led business.' (quoted in Ford, 1984).

CM & EE was seen as the priority to be reorganized on commercial lines and as such was the first BR department to adopt 'sector' management. The reorganization took place in several stages. By July 1980 the traction and rolling-stock functions were combined under one engineer and the division was restructured into three product groups (Freight, Inter-City and Suburban), as shown in Figure 23. Each group was headed by a product engineer with his own staff and supported by specialist sections for Electrical Equipment, Power Equipment, etc. There was also a Services section containing such support functions as the General Design Office and Vehicle Testing (Taylor, 1979).

Staff, including the APT Project Group, were reallocated within this new structure. The hundred and twenty 'APT posts' were dispersed among nine different sections—seventy in the General Design Office. The APT Project Group officially ceased to exist on 14 July 1980 and the APT Design Engineer was promoted to the new post of Inter-City Engineer, in charge not only of the APT but of all inter-city rolling-stock. The development work on the

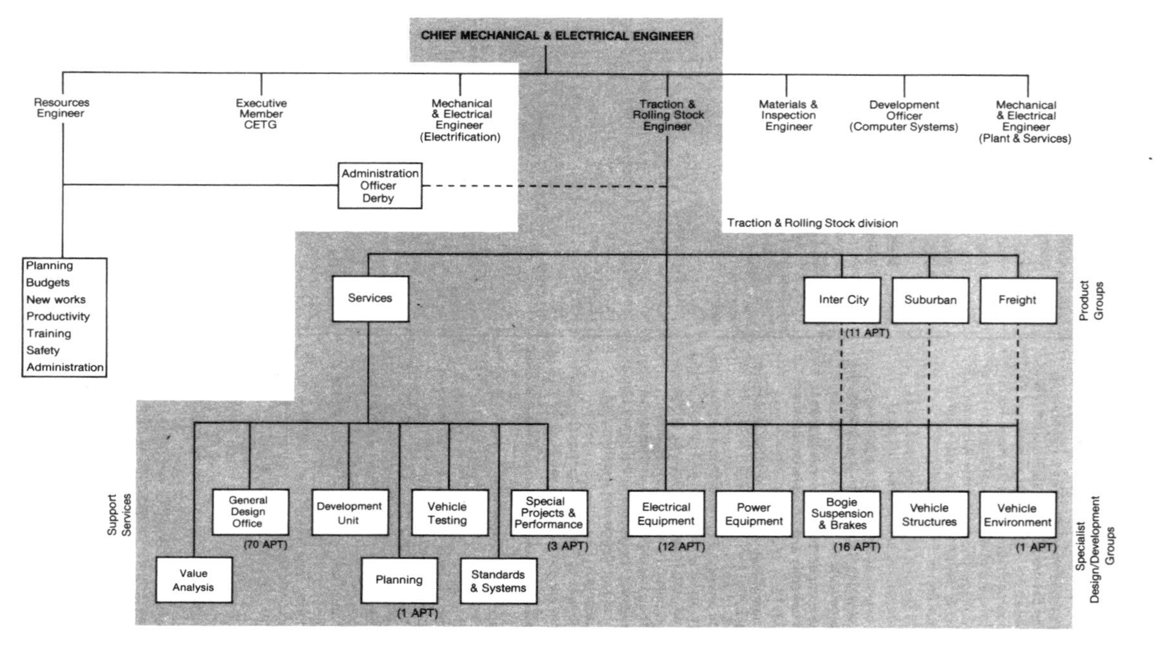

Figure 23. The CM & EE Department following the reorganization that came into effect in July 1980
Members of the APT Project Group were dispersed into nine different sections in the new structure, as indicated on the chart
Source: Potter & Roy, 1986, p. 49

APT was to continue as a project sponsored by the Inter-City sector and conducted by staff allocated to work on it by the Inter-City Engineer.

The disrupting influence of a major reorganization cannot be underrated, particularly one that is designed to reorientate the way in which an industry works. Ford (1985) summarized the CM & EE reorganization as follows: 'Accepting Gerald Fiennes' precept "when you reorganise you bleed", the M & EE Department has been haemorrhaging steadily for the past five years or so.'

The APT had bled particularly badly. Being a project team, its staff largely duplicated the skills that were already in the Traction and Rolling Stock Design Department. Equally, they were working on a project which was despised by some people who were likely to turn out to be their new bosses. So those who could left and obtained posts elsewhere which did not involve such uncomfortable conditions. This rapidly drained the project of its most capable and skilled people. Clearly the antagonisms within M & EE over the APT made the problems of departmental reorganization far worse for this than for any other project under development. Staff began to leave the APT Project Group from early 1977. Every single APT section head left in order to obtain an acceptable post either elsewhere in the railway or outside. Just as the APT was entering the crucial testing and debugging phase, key people were leaving prior to the whole team being broken up and dispersed. Just when the project required the strongest focus and greatest skill and resources, those resources were dissipated or lost.

The new organizational structure officially came into being in 1980. For the APT it had disrupted development since 1977 and continued to disrupt it while the new organization settled down.

Reorganizing the APT project

The Ford/Dain appraisal

By 1981/82 the APT was really bogged down, a fact clearly underlined by the December 1981 attempts to put it into passenger service. The BR Board approached the engineering consultancy firm of Ford and Dain Research Ltd to undertake an assessment of the technology and management of the APT project before deciding how to proceed. Because there was considerable confusion as to whether the root of the APT's problems were technical or managerial, Ford and Dain were asked to examine both aspects. They first examined the technology of the APT-P prototypes, which involved considering the APT-P specifications, test and operational reports and interviewing those involved in the project. Their overall conclusion was that the design concept of running a fast tilting train on the existing rail infrastructure was good. The engineering systems as a whole were almost right, but certain major sub-systems were too complex to achieve the levels of reliability needed and the maintenance resources available. The brakes and tilt control mechanisms were particularly identified as suffering from such problems, at least partly because their design had evolved incrementally to tackle problems as they appeared, when it may have been better to go back to first principles. However, overall, the technical assessment pronounced the APT-P to be sound in general principles and design.

Ford and Dain then turned their attention to the organization and management of the project. They were very concerned about the way in which work on the APT had become isolated within CM & EE and the fact that some engineers were undertaking work in areas in which they were very inexperienced. The fact that APT had been a project slotted into a department organized on divisional lines (first functional, then sector-based) was of particular concern. It was felt that the only way to implement a project such as APT in a large industry such as BR was to have a 'product champion' with access to a well-backed, strong engineering team. This would involve:

- The appointment of a project manager who had *total* technical and financial responsibility for the project.
- The project manager reporting directly to senior management.
- A committed team allocated work by the project manager.

Ford and Dain very much supported the project team method; however, their concept of a project team was somewhat different to the one that had previously existed for the APT and that was compatible with the restructured organization of CM & EE.

Managing innovative projects

In 1982, while Ford and Dain's report on the organization of the APT project was being completed, a new project manager was appointed. Following the Ford/Dain report, the project management that did emerge represented an integration of both the project team and functional department models. The role of the project manager was to co-ordinate work on his project by staff in different sections of M & EE and other BR departments. They were not removed from their sectors to work exclusively on APT, but the project manager was given the authority and backing to ensure that his work got done. So this effectively represented the drawing together of a project team *within* a functionally organized department. In other words, the APT project worked as if it were based in a *matrix* organization (Figure 24). This structure was considered to have overcome the main drawbacks of the original APT Project Group structure.

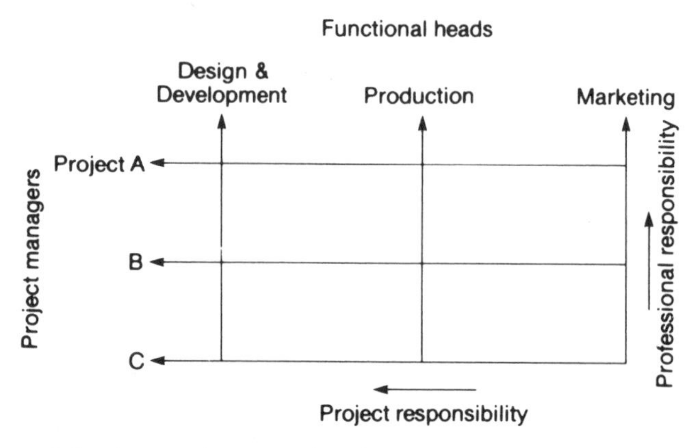

Figure 24. A matrix organization
The members of a project team are responsible operationally to their project managers, whereas they also remain members of their functional departments and responsible to the heads of those departments for the professional quality of their work and career development. The matrix organization exists only for new projects that need it, and does not replace the existing organization

One major problem with the original project organization of APT was that, because people had been exclusively appointed to the APT Project Group, the skills in that team remained static. The sort of skills needed for the different phases of development change all the time. At the beginning the bulk of the work is in engineering design, but very different skills and experience are necessary for testing and debugging. What happened with the APT Group was that the skills which were the strength of the team in the earlier stages of the project became less relevant later on. Or rather, to the design and engineering skills others should have been added. Most of the APT Project Group had little management experience outside the project itself, so although their engineering experience was extensive, they lacked ability to cope with the development and testing phases of the APT, which required co-operation and co-ordination of people from many different parts of British Rail.

Not creating an isolated team, but drawing people as necessary from within the sectors of M & EE, made it possible to change the skills in the project team to match the differing needs of work as it progressed. Equally, from a staff point of view, this matrix structure provided a settled situation, with none of the disturbances associated with the break-up of a project team once the project ended. Hence there are no problems of key people getting out once the end of the project is in sight. There is a structure which moves them smoothly on from one project to the next. Finally, although internal rivalries will always exist within any large organization, such a flexible project team structure makes the avoidance of conflict that much easier. Broadly speaking, therefore, the aim of the new management structure for APT was to combine the experience and continuity of a functional department with the direction and cohesion of a project team.

By way of comparison, this is virtually the exact management structure used for all new projects by SNCF, including the TGV. Although the technology and general approach of the TGV is inappropriate for Britain, there are real lessons to be learnt in project management that have a universal application.

The effect of this reorganization in sorting out the bugs in the APT-Ps was substantial. By early 1984 the APTs were running reliably and, from August 1984 to early 1985, they saw regular passenger service as relief trains. In December 1984 an APT covered the Euston–Glasgow run in a record 3 h 53 min, at an average of 166.4 km/h (103.4 m.p.h.), an hour and ten minutes faster than the fastest scheduled train.

The APT in transition

Today there is no reason why a fleet of APTs could not be built. But it is not to be, and the reasons for this are quite separate from the technical and managerial problems considered so far. Quite simply, in the ten years that had elapsed between drawing up the passenger specification for the APT and the successful restructuring of the project's management, market conditions had changed. In consequence, the type of trains needed had changed. The national economy, rather than booming as had been forecast, had stagnated, to be followed by a recession, with unemployment hitting record levels. In addition, the strong 'free market' philosophy of the Conservative Government elected in 1979 had led to a cut in government support to rail and the encouragement of intense competition from road coach companies. The inter-city rail market and competition between rail and other methods of travel had altered considerably. The 1973 concept of the APT was no longer relevant.

The problems of adapting a design to unanticipated changes in market conditions was mentioned with respect to the HST. With the APT, specification plans for the production version kept altering as time went by (see Figure 25). By 1981, the size of the train had been reduced and its operational speed cut, such that only one power car was now necessary.

The idea of a smaller, slower (225 km/h) APT had been raised in 1973 when studies of a single power car version were undertaken in conjunction with drawing up the commercial specification for the train. With attention focused on the twelve-to-fourteen carriage, 250 km/h double power car APT-Ps, this eleven carriage, 225 km/h version received little attention (see Figure 25c).

At that time, a range of operational speeds for APT had been considered. To begin with it was realized that even the 250 km/h potential of the APT-Ps would not be possible, owing to signalling constraints. It was accepted that until these were overcome the trains would have to be limited to 200 km/h. But even if this signalling constraint could be overcome, the case for running above 225 km/h seemed very marginal. The longest route considered for the APT was London to Glasgow, where estimated time savings between 225 km/h and 250 km/h were minimal. Conventional 160 km/h trains manage this 401-mile (645 km) run in five hours. By running at 200 km/h and being able to tilt and so curve faster, the APT could reduce this to four hours. Going up to 225 km/h knocks only a further eight minutes off this time, to 3 h 52 min, and raising the maximum speed to 250 km/h would only reduce the overall journey time by a further three minutes to 3 h 49 min. This is because curving speeds were again acting as the limiting factor on this route. Even with tilting coaches, improvements in the top speed would be unable to reduce overall journey time by very much because of the need to slow down for curves.

Hence, although the prototype APTs were built to run at a much publicized 250 km/h, there were serious doubts right from the beginning as to the commercial value of this speed. By the time production versions came to be contemplated, 250 km/h was not worthy of serious consideration. Writing in 1982, Boocock and King noted that 'savings in journey time for speeds higher than this (225 km/h) are limited on the West Coast routes. Hence, they were talking of the production of APT-S (S for 'squadron') trains consisting of a single power car with a cab at one end of a train of nine coaches plus a driving cab and luggage van (Driving Van Trailer) at the other end (Figure 5d). By reducing the size of the train to only nine coaches it is possible to use such a 'push-pull' format with the power car at one end. The propulsive force against the end coach when it is being pushed is not sufficient to

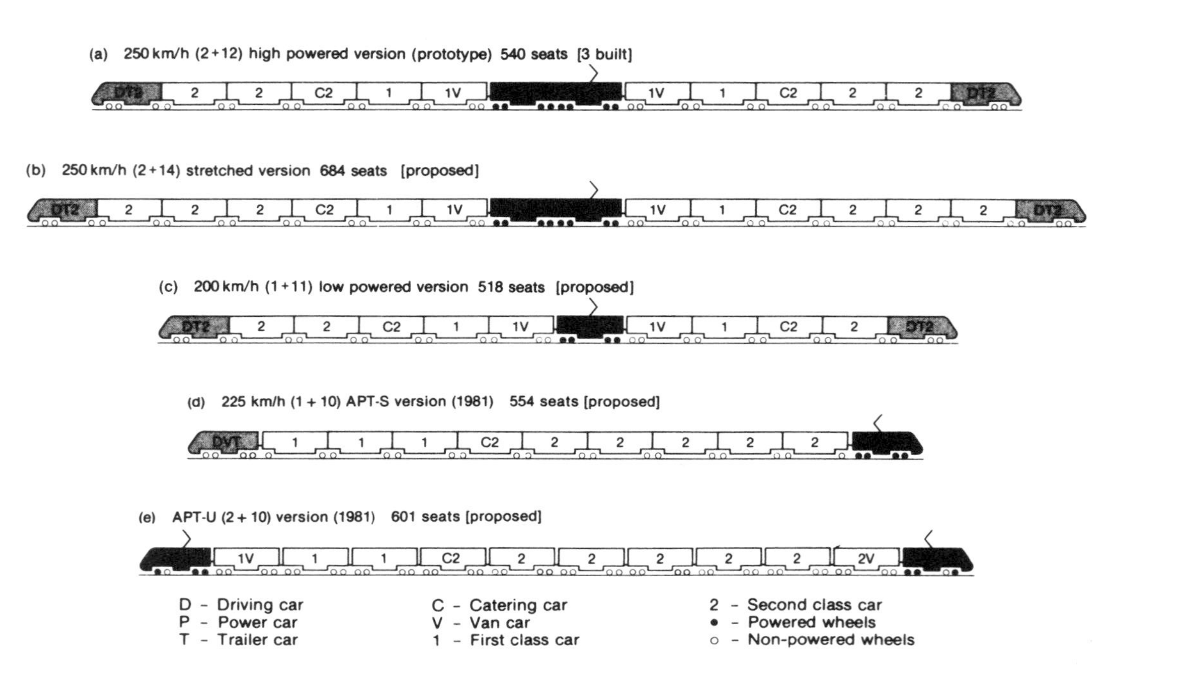
(a) 250 km/h (2+12) high powered version (prototype) 540 seats [3 built]
(b) 250 km/h (2+14) stretched version 684 seats [proposed]
(c) 200 km/h (1+11) low powered version 518 seats [proposed]
(d) 225 km/h (1 + 10) APT-S version (1981) 554 seats [proposed]
(e) APT-U (2 + 10) version (1981) 601 seats [proposed]
D - Driving car
P - Power car
T - Trailer car
C - Catering car
V - Van car
1 - First class car
2 - Second class car
● - Powered wheels
○ - Non-powered wheels

Figure 25. Changes to the specification and planned size of the APT, 1973–81
In 1973, three configurations for the production version of the APT were under consideration. These were: (a) 12 carriages & 2 power cars as in the APT-P; (b) a stretched 14-carriage version and (c) a slower (200 km/h) smaller, single power car version; (d) By 1982, the design of this smaller APT had been refined by relocating the power car to one end of the train. Called the APT-S, the disposal of the centrally located power car and the need for a second catering vehicle resulted in this smaller train, carrying 36 more people than the previous design for the small APT; (e) the APT-U design returned to a double powercar format, but dropped vehicle articulation

Source: Potter & Roy, 1986, p. 52

damage the suspension, as is the case with a larger train (this was discussed in Chapter 4 with regard to the design of the APT-P).

One criticism of the APT concept must be that if it was realized that there was such an insignificant improvement in journey time between 225 km/h and 250 km/h, why was the latter chosen as a design objective for the train? It almost seems as if the design engineers, having shown that 250 km/h was technically achievable, were unwilling to accept anything less, even though no commercial case existed for such a speed. A somewhat more sympathetic interpretation would be to emphasize the fact that the APT-Ps were prototypes, and that building prototypes to a higher performance than was necessary for actual service would ensure that any demands made of them would fall well within the capability of the technology. The TGV has run at up to 380 km/h, which SNCF claim is to show that operational speeds of 270–300 km/h are well within the technical capabilities of the train.

Was the 250 km/h design speed of the APT a case of prudent development planning or was it that, just as the leading engineers of the 1930s had used their power to get the railways to invest in fast steam traction rather than the more suitable diesels, so the engineers of the 1970s had got commercial management to accept a design speed that had no commercial viability? For the moment this question must remain unanswered. It is probably not constructive to even attempt an answer as it is now no longer relevant. One major goal of the restructuring of BR into sector management has been to make the railways 'commercially led'. It has been to provide a clear structure for evaluating all proposals in terms of precise, commercially justified, objectives. A situation where the operational and management side of BR is unable to evaluate a proposal from research and engineering should never occur again.

A commercial limit to train speeds?

It now seems that the true limitations to fast train speeds are not technical but commercial. Even for the longest inter-city route in Britain, above 225 km/h the increase in passenger revenue cannot match the increased cost. Journey times are not reduced to any significant degree, whereas fuel and maintenance costs continue to rise (see Figure 26). British Rail commercial models have established that, for long-distance traffic in Britain, it is not commercially viable to go faster than 225 km/h and in many cases 200 km/h seems more appropriate. Thus 225 km/h is the design speed for the trains that are being developed from the APT project.

Speed is now no longer as great a priority for BR as it was a decade ago. Models of passenger demand, as discussed in Chapter 6, have shown that other factors can significantly influence passenger traffic. In the 1984 Inter-City Strategy document, *InterCity into Profit*, speed received remarkably scanty attention:

> With many services already operating at 125 m.p.h., increased speeds are not an immediate priority, but steps will be taken to increase top speeds and improve timings on the main business trains between London, the Midlands, the North-West and Scotland. Where top speeds are currently in the 100/110 m.p.h. range, these will rise to 125 m.p.h. as new locomotives come into service in the late 1980s. In the 1990s, it may be that further improvements in electric locomotive design, track and signalling will permit top speeds of 140 m.p.h., twice the speed permitted for the car or coach on the motorway. [British Railways Board, 1984, p. 15.]

Perhaps BR should not be unduly criticized for developing a train with a performance that is somewhat too high. The mistake is negligible compared to those made in some contemporary projects such as Concorde—a techni-

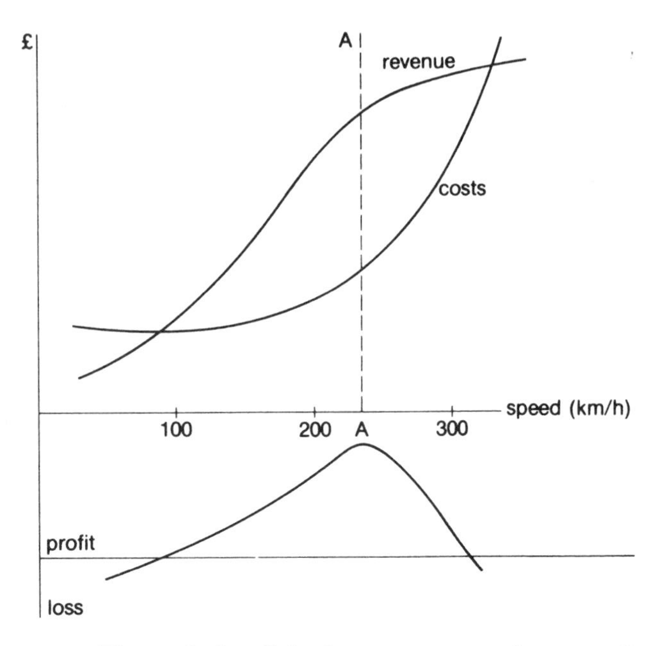

Figure 26. The relationship between speed, operational costs and passenger revenue
Beyond a certain point, the costs of additional speed exceed the revenue that lower journey times generate. This represents the notion of maximum 'commercial' speed. This is likely to vary given different markets and technologies.
Source: Potter & Roy, 1986, p. 53

cally successful aircraft whose specification had little commercial input and whose speed generated insufficient traffic to cover its costs. Marketing models and information were not sufficiently sophisticated at the time to foresee such aspects and, under the circumstances, British Rail did considerably better than most.

If anything, BR's lack of criticism of the APT's design speed can be more readily understood since, through the 1960s and 1970s, several other nations were investing vast sums in the development of trains of up to a 300 km/h capability. The TGV in France is perhaps the most obvious example, followed by Germany's 250 km/h Inter-City Experimental (ICE) train. But the distances involved are different to those in Britain; there is more long-distance traffic so that, if new track is being built anyway, the cost of high speeds can be justified. Even so, the TGV is not authorized to run at the 300 km/h it is capable of simply because of the fuel costs incurred and in practice it does not run at its much publicized 270 km/h for extended periods for exactly the same reason. Indeed, in recent years, the French have been putting a lot of work into 'coasting' fast trains downhill so as to save energy costs!

Furthermore, there was no commercial evaluation of the raising of the maximum speed from 270 km/h to 300 km/h for the TGV Atlantique line. This occurred for purely technical reasons, resulting from an improved engine design. Since it was technically possible, SNCF adopted 300 km/h running. Since on this route the new line will comprise a small proportion of most journeys, the additional time-saving between 270 and 300 km/h will be insignificant. When I asked a senior executive at SNCF what time-saving this represented, he admitted that this had not been estimated and guessed that it would be 'at most', five minutes. This lack of commercial input into French rail planning stands in marked contrast to the strong French management structure. The greatest value in raising the TGV's top speed is probably in terms of publicity and image (which, again, SNCF traffic models do not take into account!). Also, the political strength this speed capability carries is valued by the French, particularly as joint venture train projects are now emerging. This aspect is developed further in Chapters 8 and 9.

Rather than comparing the British studies with SNCF, where there are clearly far from commercial influences on design speeds, a more valid comparison may be the Swedish Railways tilting train project. This has drawn very much on APT experience and will operate on routes with a similar distribution of population to that envisaged for the APT. This has a planned top speed of 200 km/h. Three prototypes are currently under construction and a fleet of fifty are due to enter service from 1990 to 1994.

So, by the early 1980s, attention was focusing on the smaller, slower single power car version of the APT. But in the wake of the 1982/83 management reforms. The Ford/Dain review and the strongly commercial reorganization of BR as a whole, a very different design concept for the APT was to emerge. Indeed, the train to be developed from the APT, while sharing some of its technology, represents a very different design concept.

8 The successors to the Advanced Passenger Train

The Adaptable Passenger Train

With the fall in expected passenger demand on the West Coast main line, the design concept for the production version of the APT had, by 1981, focused down upon the smaller, slower version of the train. The APT-S design consisted of a single, almost locomotive-like power car at one end with nine passenger coaches and a driving van trailer (DVT) at the opposite end to the power car. This was soon dropped in favour of a non-articulated design with power cars at either end, called APT-U (Figure 25e). The basic concept of a fixed formation, dedicated fast passenger train remained. But in 1983, following the Ford/Dain report, together with an associated review of the design requirements for fast passenger trains, the whole notion of developing the existing APT design into a production train was abandoned. The experience of the APT project is now being used to produce a train of a significantly different design. The APT-P, instead of being a prototype for a production fleet, was destined to become a development testbed for technologies to be incorporated into another design of train. The passenger trials ceased in 1985, leaving only one operational APT-P for developmental work. This was redesignated APT-D to emphasize its change of function.

There are three main reasons for this change. The first is that the APT was very much a 'lean' design (some problems of which were discussed with respect to the HST in Chapter 6). Although some aspects, such as its top speed and the centrally positioned power cars, paid rather more attention to technical than operational requirements, overall the APT was a well-integrated design concept. But times change and markets change and, as had been found with the HST, one design concept tailored carefully to fulfil a particular job can result in difficulties if that job alters. It was feared that the articulated, fixed formation APT could prove to be too inflexible to future changes in rail's passenger market. A more 'robust' adaptable design was called for than the 'lean' APT concept.

This approach was reinforced by the far more rigorous and integrated methods now used to evaluate individual investments in BR, which increasingly sought to find the best solution for a business sector as a whole, rather than within the narrow confines of a particular project. The plan for the future of the APT was therefore integrated into wider plans for BR's traction and coaching stock. This comprehensive systems-approach concluded that it would be more economical to develop the power car as a conventional separate locomotive. As such, it could be used to haul freight, parcels and sleeper services, which it could not do if it were permanently part of a fixed formation train set. In actual practice, the specification of the locomotive focused down on a dual 'day-time fast train/night-time sleeper' role, with a utilization averaging 420,000 km a year. The APT power cars are therefore being treated as prototypes for the Class 91 'Electra' locomotive and the carriages used to help develop the design for a new generation of 'Mk4' coaches (see Figure 27). This design approach stands in strong contrast to that of the TGV, for the French have retained their fixed-formation, articulated sets for the TGV Atlantique stock.

The same comprehensive approach also concluded that it would be more efficient to standardize on one basic design for passenger stock for both the East and West Coast main lines. The electrification of the East Coast main line originally envisaged the use of relatively conventional electric stock to replace the diesel High Speed Train. This

Figure 27. Artist's impression of a Class 91 'Electra'-hauled train
Source: GEC

Note the way in which the design of the rear cab and the livery give the impression of a fixed formation train. The configuration of the InterCity 225 will consist of 11 Mk4 coaches, the Class 91 locomotive and a driving van trailer, giving a seating capacity of about 700

included the use of Mk3 coaches (as used in the HST) and 200 km/h Class 89 locomotives, a prototype of which was ordered from Brush. The performance specification of the Class 89 represented a modest improvement over that achieved by existing electric locomotives (e.g. the 175 km/h Class 87), so overall the new stock was essentially planned to be little more than an electric replacement for that currently in use.

However, in 1985, BR decided to integrate plans for the East and West Coast lines and use basically the same locomotives and carriages. This involved a change in specification for the East Coast main line to Class 91 locomotives and a non-tilting version of the Mk4 coach. With the first stage of the electrification (from London to Leeds) due to be completed by October 1989, the time-scale is tight, so development effort is being concentrated on the Electra locomotive and the non-tilting Mk4 coaches. The Class 89 has been reduced to the status of an 'insurance policy' should there be delays in the Class 91's development. Within a couple of years, the whole focus of the APT/Electra project was thus shifted from the development of a specialist tilting train for the West Coast main line into a rapidly developed non-tilting train for the East Coast main line, from which a tilting version may then be built.

Contracts for the construction of the first batch of 31 Electra locomotives were let in February 1986 to GEC Transportation Projects, who have subcontracted a substantial amount of work, including the final assembly of the locomotives, to BREL. The APT design and development experience is being used to aid the rapid development of this train. Because of this and the existence of the APT-D, no prototype Electra is considered necessary. The first train is due to be delivered to BR in late 1987, with trials commencing on the West Coast main line in 1988. The first ten trains are planned to inaugurate the London–Leeds electric service in October 1989, just over three and a half years from the letting of the contract. This is almost exactly the same time-scale as was achieved with the HST.

But although the development of the Electra benefits from the general experience of the APT project, very few design features of the APT are directly incorporated into its successor. The main transfer between the projects is in terms of theoretical understanding and drawing on the practical experience of the APT design process.

GEC's mechanical engineers achieved low track forces

and an unsprung mass of 1.7 tonnes using a different design of body-mounted traction motors from that used in the APT. These are hung below the body, producing a low centre of gravity (minimizing body roll and pantograph movement), and permit a conventional layout inside the locomotive, allowing ease of maintenance. The final drive to the wheel-set uses a flexible drive developed by GEC which is simpler and more efficient than the final drive on the APT. With a continuous power rating of 4.53 MW, although 225 km/h is the maximum service speed envisaged, the Electra will actually be capable of 240 km/h—only a shade slower than the APT. This high power output is a consequence of the locomotive's dual function. The 4.53 MW are necessary for it to be able to haul 830-tonne sleepers and to meet emergency operating conditions.

The one area of direct APT influence is in the streamlined cab design, although, as a locomotive, the Electra requires buffers. However, one unique feature of the Electra locomotive is its asymmetrical design. Like all electric locomotives, it can be driven from either end, but the driving cab will be streamlined at one end only. There are two reasons for this, one technical and one commercial. The technical reason is that if the cabs at both ends were streamlined there would be a gap between the locomotive and the coaches, which at high speeds would cause unstable turbulence and drag. The second reason is that, although a fixed-formation train may be operationally inflexible, it has a very strong passenger image, and the Inter-City sector management of BR did not want to see this lost. Hence, having a cab flush against the adjacent coach produces an impression of a fixed-formation train, yet retains the operational flexibility of a locomotive. The name Inter-City 225 is proposed for Electra-hauled 200 to 225 km/h passenger trains in order to link its image with the fixed-formation Inter-City 125 HST. When driven from the bluff end, the locomotive will be limited to

175 km/h, which is hardly a constraint since sleeper and any other uses operate at below this speed.

For the coaches, although these will be used in a fixed-formation rake, articulated bogies have been dropped. One element emphasized in the Ford/Dain report was that it was only possible to innovate as fast as the *organization* could manage the innovation process. A concentration on 'key' design innovations and the simplification of others was advocated in order to reduce the amount of managerial effort and co-ordination needed. There had been problems with the articulated bogies of the APT and its lightweight aluminium carriages, so it was decided to opt for a more conventional, non-articulated design and to concentrate development effort on the 'core' innovation of tilting a coach reliably and adapting the power cars into a lightweight locomotive. This obviously imposed a weight penalty but, with the maximum operational speed lowered to 225 km/h, this had become less of a concern than with the 250 km/h APT.

In effect, with the shift in the initial development of the Electra/Mk4 coaches to the East Coast main line, the coaches will first be developed in a non-tilting version. For this relatively straight line, the use of tilting coaches would only reduce the overall London–Edinburgh journey time by five minutes, which is insufficient to make their use commercially worthwhile. As such, the development process of the APT's successor has not only been simplified, but broken into two distinct stages. Following the introduction of the first Inter-City 225s, using non-tilting Mk4 coaches on the London to Leeds route in 1989, the tilting version will be developed for service on the West Coast main line in the early 1990s.

The Mk4 coach design is much simpler than the lightweight, aluminium coaches of the APT. They are to be built in steel, and the tilting version will only tilt at 6°, not 9° as in the APT. The use of a common bodyshell for both tilting and non-tilting coaches was made easier by restricting the tilt to 6°. A wider, more spacious coach than on the APT is possible, while keeping within the loading gauge. However, although the InterCity 225 will not tilt by as much as the APT, it is still planned to take curves at the same speed as the APT. Following passenger tests, it was concluded that 9° of tilt for 9° of cant deficiency was unnecessarily generous, so whereas the APT compensated for all lateral forces, passengers in the Mk4 coach on an InterCity 225 will feel the residual 3°. This is still better than the 4.25° of cant deficiency on existing non-tilting trains. Total compensation for cant deficiency would have cost a lot more, by requiring the development of two totally different coach designs, with only a marginal increase in comfort and commercial attraction.

The residual 3° of cant deficiency that will be felt by passengers on curves may actually be an advantage of the InterCity 225 design. One non-technical problem associated with the APT-P was that passengers were found to be more prone to travel sickness than in conventional trains. When travelling in the APT-P on one of its relief runs in December 1984 I found myself feeling quite queasy, although I have never suffered from travel sickness in a train before. A number of other passengers were also affected. There are two factors believed to cause this. One is technical–the poor ride of the articulated bogies and a high level of vibration in the coaches. Plans to eliminate these problems are considered below. The second factor is physiological. The 9° tilt on the APT is actually too good, in that it compensates for all lateral g-forces. People travelling in the APT have no sensation of cornering when their eyes tell them it is happening. This, coupled with the occasional rapid tilt around steep reverse curves, causes some passengers to suffer from 'tilt nausea'. With a residual 3° of lateral g-force, people will be aware of curving and hopefully this source of travel sickness will be eliminated.

The tilt mechanism on the APT-Ps is now working

reliably. This can indeed be taken as quite an achievement. Although BR has been working on this since 1967, other railways have also been working on actively tilting coaches and none have as yet succeeded in developing a sufficiently robust, commercially viable design for tilting coach stock. The LRC (Light Rapid Comfortable) design in Canada/United States has had continual problems which have not been resolved and the Swedish railways design (modelled very much on the APT) is still a long way from passenger service. Italy's *Pendolino*, capable of an 8° tilt, is the closest to passenger service. Even so, development work on this has been even more protracted than on the APT-P. The prototype was built in 1975, with it taking over ten years to develop it sufficiently to contemplate a production version.

The tilt mechanism, although now working well on the APT-P, is still considered insufficiently robust for use on the Mk4 coach and a number of options are being followed up in order to make the design more robust and tolerant of faults. One example of this that is under consideration is a new sensor to detect curves and provide the signal to the tilt mechanism without interference from other electrical equipment.

The approach to tilting coaches for the Mk4 carriages displays a much more thorough analysis than was the case with the APT. The 9° tilt was taken for granted throughout the APT programme. The costs involved in producing non-standard stock were realized, but seen as necessary to achieve the train's speed target. In practice this was too simple an analysis, the present arrangement for 6° tilt emerging out of the new management and project anlaysis structures. What is perhaps more surprising is that a BR study is under way examining whether to build a tilting version *at all*. The study is linked to whether BR really considers it can capture enough business traffic with the tilting Mk4 coaches to make their development worthwhile. This has arisen out of the erosion of business traffic on the West Coast main line in the absence of any major speed improvements in the last decade. As such it is a reflection of the absence of tilting coaches, rather than of the crucial original decision to make the APT a tilting train.

The Mk4 coach of the InterCity 225 is expected to be built in steel. The reason for this relates to ride quality. The lightweight coaches of the APT-P had a poor ride, largely because the suspension had to be softer than for heavier coaches, which meant there was a greater risk of them being affected by general vibration and any stiffness in linkages or other components. All objects have a natural vibrational frequency, so that if a source of vibration of the same frequency is near, then the object itself will vibrate. Rather than attempting what could have turned out to be a long and complex development process on the lightweight APT coach, a much simpler approach was taken. This involved just taking the good riding quality of the Mk3 coach as a specification to be achieved by the Mk4. The natural frequency of the Mk3 coach was 10 Hz (cycles per second), so this became the target to achieve for the Mk4 coach.

Computer studies were used to examine what sort of coach designs could achieve 10 Hz. It had already been decided to use 23 metre-long coaches to provide maximum passenger capacity and, given this and the simpler, more conventional, development methods adopted, it was found that an aluminium coach meeting the specification would actually weigh more than a steel design simply because of the amount of material needed to achieve a 10 Hz frequency. Overall, the Mk4 coaches for the InterCity 225 will be of a much simpler, more conventional design than the APT coaches. Athough the APT-D is being used to help design the Mk4 coach, it will incorporate very little of the original APT design. The APT legacy in more in terms of design methodology than hardware.

Table 3 Weight and capacity of British Rail coaching-stock

Coach type	Date of introduction	Max. seating capacity	Structure Kg	Mass per seat (Kg)
Mk 1	1952	64	9,810	153.3
Mk 2	1964	64	7,630	119.2
Mk 3	1975	72	8,300	115.3
APT-P	(1981)	72	5,720	79.4
Mk 4	(1989)	80	8,500	106.3

Sources: Ledsome, 1981; Boocock, 1985.

In terms of construction method and weight, the Mk4 coach is closer to the Mk3 coach than the APT trailer car (see Table 3).

Although designed to be capable of 225 km/h, the InterCity 225 will initially be introduced at 200 km/h, owing to the limitations of existing signalling. Once in service, British Rail will then be in a better position to judge whether it is commercially viable to invest in new signalling to raise speeds to 225 km/h. Models of elasticity of demand with speed in the 200 km/h to 270 km/h range have not proved reliable and, for this reason, BR are seeking to have some actual operating experience before deciding to opt for any higher speeds. The extent to which BR is keeping its options open can be seen in the electrification scheme on the East Coast main line, where the overhead wires and supports are designed for 225 km/h, but can easily be modified for trains running at 270 km/h—even faster than the APT!

To run at 225 km/h would require a longer braking distance than the present four-phase block signalling permits. As was noted in Chapter 5, the idea of adding a fifth signalling aspect to give additional braking distance was simply shelved and the APT restricted to 200 km/h. The method currently favoured to provide this 'fifth aspect' is a development of the electronic line speed advice system which was installed in the APT. The conventional way of advising drivers what speed is permitted on any section of track is to display the line speed on trackside boards. Because of its lightweight design, the APT can go faster than this, so a method had to be devised to inform the driver of the line speed for the APT. This was done electronically, by mounting transponders in the track which supplied an electronic signal to equipment mounted on the train. A cab display then informed the driver of the APT's maximum line speed for that section of track.

If this transponder system were modified to link into the signals one block ahead, then advance warning could be given to a signal set at 'caution' (double amber) and the train slowed down to 200 km/h so that it could then stop within the normal braking distance. This would be very cost-effective. Only trains that needed an additional braking distance would have such equipment fitted and there would be no requirement to undertake costly modifications to existing signalling equipment. Work on developing this signalling method is currently under way. Its use will depend on whether there is an economic case for operating at above 200 km/h.

Overall, the changes to the specification of the APT have produced a substantially different train. Instead of a dedicated, fixed-formation train, what has emerged is an adaptable high-performance locomotive and a set of reasonably lightweight coaches which can be used for a wide variety of purposes, one of which is a 225 km/h passenger train. The design concept is less innovative and less specialist. The 'robust' rather than 'lean' approach means that it can be adapted to serve a wide range of railway operations and should be considerably more flexible and responsive than either the HST or APT to unpredictable changes in market conditions.

Managing the InterCity 225 project

As can be gathered from the above section, the design of the Class 91/Mk4 Coach InterCity 225 concentrates on the core developments needed to run a fast train on existing curved track. The development process has been considerably simplified by reducing the train's top speed and by dropping a number of weight-saving and specialist features such as articulated bogies and light aliminium coaches. This is primarily in order to make the project more manageable.

In addition to making the development of the InterCity 225 simpler to manage, the management structure of the project has taken on board the lessons learnt from the APT. The management reforms for the APT, as discussed in Chapter 7, have been retained for the InterCity 225. There is an InterCity 225 Project Engineer who is responsible for assigning people to work on various aspects of the train and for the overall co-ordination of the project. Within the M & EE Department, which underwent further reorganization after 1980, the project is based in the InterCity Business Group, but managed on a 'matrix' basis, with people from relevant areas of the Traction and Rolling Stock Division and elsewhere being drawn in to work on the project as necessary (See Figure 28).

A final area in which the management of the InterCity 225 project has been simplified is in the relationship between British Rail and the rail industry. The role of British Rail in design and development is far less dominant than it was. The in-house design work on the Class 91 locomotive and Mk4 coaches has been undertaken in order to provide *specifications* for tenders from manufacturers. Although there will be close co-operation between the successfully tendering companies and the M & EE Department, responsibility will be with the manufacturing companies and not with British Rail. This shift in responsibilities between BR

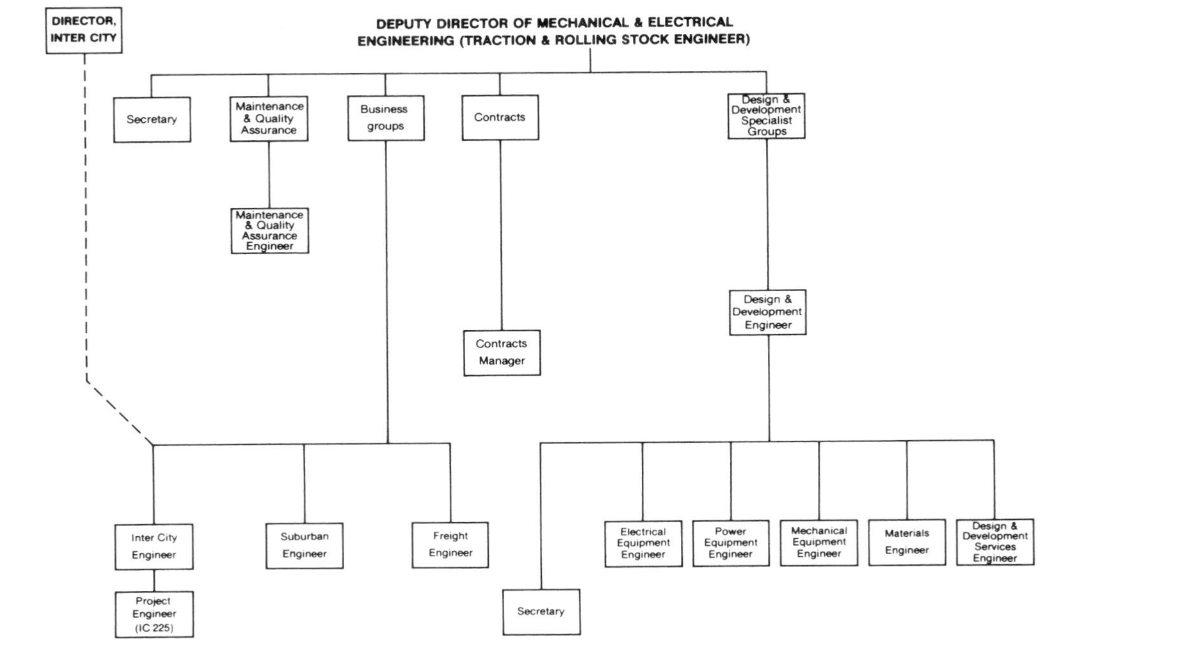

Figure 28. Organization of the Traction and Rolling Stock Division within the Mechanical and Electrical Engineering Department in 1984
The InterCity 225 project is managed on a matrix basis from a location in the InterCity Business Group under the overall control of the Director, InterCity

Source: Potter & Roy, 1o86, p. 58 (after British Rail)

and Britain's railway industry is very significant. A government inquiry into public sector research and development (ACARD, 1980) had recommended that, in order to improve their export competitiveness, state-owned companies should rely less on in-house research and more on their suppliers and other contractors to develop new products. Although in-house research is very necessary for the development of basic innovatory ideas (such as BR Research's development of the APT-E), the greater involvement of supply companies in taking over and developing such designs widens and stimulates a greater field of expertise.

Clearly the ACARD report was welcomed by the Conservative Government who were keen on such 'privatization' but, quite aside from short-term political dogma, it is likely that British Rail would have done this in any case. They considered themselves to be too dominant in the design field which they felt was not encouraging innovation in their supply industries. In this particular case 'privatization' was not forced on BR by government but was more a label attached to a process that was already under way. In other areas of BR's activities, privatization has been far from welcome, and has only occurred with much reluctance. However, this new procurement policy has organizational implications that BR is only now beginning to understand. Firstly, it means that private supply companies will be tendering against BR's own subsidiary, BREL, so BREL itself has had to be split into 'maintenance' and 'production' wings, with maintenance continuing to work very closely with other BR departments, while production has more or less the status of an outside company. Secondly, many of the duties previously undertaken by M & EE staff are now undertaken by the successful tendering company. This realignment of staff posts may well result in similar problems of personal insecurity, as did the major M & EE reorganization in the late 1970s.

The Euro Train

In January 1986 the authorization of the Channel Tunnel project was announced. As had been widely expected, the successful contender was the Channel Tunnel Group's scheme for a twin railway tunnel permitting the operation of a rail 'shuttle' service for road traffic combined with through, long-distance passenger and freight trains. The linking of the British Rail network to that of Europe requires the development of a new generation of fast international passenger trains. The market-based studies which showed 225 km/h to be the commercial limit to speed in Britain were based upon Anglo–Scottish flows on the East and West Coast main lines. Over the London to Paris, Brussels, north Germany and other European routes, very different markets are involved, which will include substantial running over the TGV Nord and the German *Neubaustrecken* lines. Given this linking of BR's network to the emerging European network of purpose-built high-speed lines, the basic design specification agreed for the Joint Venture High Speed Train to operate Channel Tunnel services includes a 300 km/h capability.

The development of this train is likely to involve an international consortium of railway manufacturers and within months of the authorization of the Channel Tunnel project rival bids for project leadership were emerging in Britain and France. SNCF was openly referring to the train as the 'Channel Tunnel TGV', while a consortium of Britain's main traction and rolling-stock manufacturers was rapidly formed and in the spring of 1986 they presented their own plans for the Euro Train.

The Euro Train is a direct adaptation of the Class 91 Electra and Mk4 coaches for 300 km/h operations. This involves two Electra-derived power cars at each end of a rake of aerodynamically enhanced Mk4 coaches, making up a fixed-formation train seating 840 passengers. The power cars would have to be multi-voltage, to be able to operate on a 750 V third rail was well as the 3 kV and 25 kV overhead line systems. There is no proposal to use tilt on these trains, the plans for the purpose-built TGV and *Neubaustrecken* lines rendering this unnecessary. But, where European services include substantial runs over existing lines, the use of tilting carriages would produce significant time-savings. Indeed, an Electra running over existing track at 240 km/h could well match a TGV running at 300 km/h over special track but limited to 200 km/h elsewhere.

But Europolitics will undoubtedly play the major role in the eventual outcome of both the lines and stock for future cross-Channel trains. The eventual design of the Euro Train must be expected to involve a number of European railway manufacturers in a joint design. The purely British Class 91/Mk4 coach derived design, at least as currently envisaged, is unlikely to be built. But, as Roger Ford points out:

> while this may be 'realpolitik', . . . the starting point for European collaboration must be 'we can do it on our own' rather than 'let's do it together'. Experience has shown that this opening gambit has served the French aircraft industry, for example, very well at the expense of the sometimes naíve Brits.

This will involve yet another factor in the innovative process, for as Mowery and Rosenberg (1985, p. 95) point out, 'Tasks of management, cost control and design that are at best challenging in a single firm can become overwhelming in a multinational consortium.' They were examining commercial aircraft projects, where multinational collaboration is now well established. The railways are not used to multinational collaborative projects and the Joint Venture High Speed Train is likely to be the first major example. Mowery and Rosenberg suggest that the multinational collaboration is most successful where there is 'an already clearly established working relationship between junior and senior partners' and where 'joint ventures are also working with relatively stable technologies and desired characteristics' (p. 96).

Although the technology of the Joint Venture High Speed Train may be seen as stable and there is likely to be reasonable unity in the desired performance characteristics, a totally unknown factor is the managerial and working relationships between the French, British (and possibly German, Dutch and Belgium) railway industries. The initial moves suggest that there is going to be quite a tussle for project leadership and, whatever happens, the train's design is bound to require the use of components from several manufacturers in up to five countries.

With the APT, the major problem was in the working relationship between the research-based project team that conceived the train and the organization upon which it depended for the train's successful development. There is most certainly a lesson here for the Joint Venture High Speed Train, for if this project stirs up similar antagonisms between the companies and groups involved in its development there is a real danger of a repetition of the APT saga on a much larger scale.

9 Comparative experience

This book has purposefully concentrated on a limited number of case studies in fast passenger train design and development: the Shinkansen, the TGV, HST and APT. Other fast train projects have featured, but the focus has very much been on these four projects and, in particular, on the APT/InterCity 225 saga. The factors involved in determining the form and extent of innovations necessary are such that it is preferable to focus down on a limited number of case studies rather than venturing over a wide number of examples. However, before attempting to summarize this experience and pull out some general conclusions on the innovative process, it is worthwhile examining some parallels and contrasts with other fast train/ground transport projects currently under development.

Firstly, it is appropriate to return to the Shinkansen. The original Tokaido line took traditional railway techniques to their operational limits in order to achieve 210 km/h running. The success of this line led, as was noted in Chapter 3, to plans for a whole national Shinkansen network. The rationale for the Shinkansen shifted from that of a one-off project through to a national programme designed to help alleviate Japan's unbalanced economic and urban growth.

Three new Shinkansen lines were built in the 1970s: the Tohoku line from Tokyo to Morioka and the Joetsu line from Tokyo to Niigata, together with the extension of the Tokaido line (the Sanyo line) to Hakata in the south of the country (see Figure 9). Although they incorporated some technical advances made in the 1960s and 1970s, the design of these three lines used basically the same approach as was pioneered in the original Tokaido line. This involved building totally new elevated lines from city centre to city centre on very expensive, heavily engineered track. As Wickens (1985) notes, despite the high infrastructure costs, the original Tokaido line is:

> commercially successful because it carries in the region of 120 million passengers a year. With that number of passengers it is possible to build an extremely expensive infrastructure and still make money. The operating ratio is about 60 per cent. It really is unique and I think it is actually unique in Japan, because I do not think the other lines that are being built will achieve the same degree of commercial success.

In actual fact, the Tohoku Shinkansen carries around twenty million passengers a year and the Joetsu Shinkansen thirteen and a half million. This places them in the same range as the TGV projects, but although passenger demand is the same, the track construction method and geography of the areas these lines serve mean that capital costs are vastly higher.

But by 1982, when these lines opened, the tide had already turned against the Shinkansen. From the mid-1970s, concern had mounted over the rising deficit of Japanese National Railways. The Tokaido Shinkansen may have been profitable, but general overmanning coupled with a political insistence that fares be kept low and a lightly used rural service maintained, meant that JNR was moving deeper and deeper into the red. In 1977 the government ordered JNR to increase fares by a staggering 50 per cent, regardless of route or market sensitivity. Shinkansen traffic slumped by 15 per cent overnight, largely to the airline's gain.

Added to this general atmosphere of cost-cutting came the late 1970s' recession plus a public disenchantment with the Shinkansen itself. The latter, which may seem surprising given the pride in which all other fast train projects are held throughout the world, was due to a blend of specific political factors aggravated by the design concept of the Shinkansen itself. As Freeman Allen (1978, p. 92) noted, this souring of public opinion was:

> sparked off by growing complaints of noise from people living near the new railways. In designing the New Tokaido Line, JNR engineers had taken some care to limit noise and vibration to levels no worse than those experienced in the proximity of the 3ft 6in-gauge system, but they had not foreseen the effects of the intensive service which rising traffic demand quickly generated. By now Japan, which had previously tolerated the foulest industrial pollution, probably, of any industrial country, was on a highly emotive ecological kick. Self-seeking politicos pounced on the alleged environmental disturbance of the Shinkansen as a symbolic issue, public opinion was roused and controversy erupted over much of the projected new Shinkansen routes.

The eventual upshot was that new noise regulations resulted in the most expensive railway track in the world becoming even more expensive. The sum of £200 m was spent on noise abatement measures on the existing Tokaido line alone and the design of all other new lines was modified to reduce noise intrusion, even to the extent of burying them underground in some places.

Although by far the majority of the environmental concern for the Shinkansen was politically generated, the design concept did involve serious environmental and economic problems. Since the Shinkansen was built to a different gauge than other lines in Japan, to reach Tokyo's city centre it required a new link to be carved through metropolitan Tokyo for the Tohoku and Joetsu lines. The cost and protracted wranglings with landowners and residents delayed the opening of these lines from 1977 to

1982, and even then, the final link to Tokyo' city-centre Ueno station was not opened until March 1985.

So, while Japan's government was, on the one hand, becoming far more cost-conscious about the railways, 'environmental' design requirements for Shinkansen lines were pushing up the costs of an already expensive design concept. Although still keen on the use of the Shinkansen, by 1978 the government department responsible for regional development was pushing for a cutback in the planned network of lines. In fact, although route surveys for three more lines have been undertaken and newly designed rolling-stock capable of 260 km/h is under development, no further Shinkansen building has been authorized.

All the most heavily-used routes are now served by existing Shinkansen lines, and of these, only the original Tokaido line is profitable. On the Tohoku and Joetsu Shinkansen, revenue meets operating costs, but only makes a small contribution to their heavy capital investment. With the government less committed to the regional development function of the Shinkansen than it was when these lines were authorized, JNR is seeking to cut the cost of high-speed lines. An approach incorporating some TGV-type elements is suggested, using steep gradients and smaller-diameter trains in order to cut tunnelling costs.

But the simple truth of the matter is that the really big traffic-generating routes are already served by existing Shinkansen lines. Any further expansion of the network will not be commercially viable and will thus depend on national policy objectives. The regional development goals of the Shinkansen remain valid, especially since the planned extensions involve the underdeveloped islands of Hokkaido and Kyushu. But national policy objectives have shifted towards the separation of railway policy from other areas of state planning. The 'systems' approach to state intervention in transport planning has weakened in the face of monetary economics. Plans are currently under way to denationalize Japanese National Railways, breaking it up into six regional companies plus a freight concern. JNR is the oldest nationalized railway system in the world, and these plans are bitterly opposed. Until the future of Japanese railways as whole settles down, it is unlikely that any further Shinkansen lines will be built. If the railways then operate on a purely commercial basis, there is no prospect at all of the Shinkansen being expanded.

This situation is epitomized by the £2,700 m Seikan Tunnel which connects Japan's main island of Honshu to Hokkaido beneath the stormy Tsugaru Straits. Work on it began in 1964, but with the expansion of the Shinkansen lines in the early 1970s, the project very much became part of the extension of the Tohoku line north from Morioka, through the tunnel and on to Sapporo, Hokkaido's largest city. Tokyo to Sapporo would then involve a journey time of only five and a half hours, roughly the same city centre-to-city centre time as by air.

The 53.9 km (33.4 miles) long tunnel was completed in 1985, but the end of the Shinkansen line from Tokyo is still over 150 km away, at Morioka. From there to the tunnel runs a tortuous narrow gauge line, over which standard gauge Shinkansen trains cannot run. Narrow gauge tracks are being laid in the tunnel, which is expected to open in 1988. For the time being only narrow-gauge passenger and road vehicle shuttle trains will be able to use the tunnel. Getting to Tokyo will involve changing trains and a very long journey on a slow, narrow gauge line.

Writing in 1978, Freeman Allen said that it was 'sad that a concept which so invincibly demonstrated the validity of a dedicated high-speed passenger railway custom-built to commercial and social demands of the late 20th Century should latterly have been humbled by parish-pump politicking.' Since then, the 'parish-pump politicking' has grown considerably worse, but behind it is the realization that the Shinkansen approach really has a remarkably limited application. The whole design concept was so specialist that it was unable to adapt to political and economic situations other than those prevailing when it was conceived. The Shinkansen may have presented a vision for the future of inter-city railways world-wide, but in terms of method and technology it has little to say to the world today.

Using new rail links to upgrade large networks

The TGV-style approach of upgrading a whole railway network by the building of a high-quality new link is one that has a far wider application than trying to build an entirely new network from scratch. Whatever method of evaluation is used, be it strictly financial, cost-benefit, or national–strategic, the cost-effectiveness of the TGV approach compared to that of the Shinkansen cannot be denied. Unless traffic flows of around fifty million passengers per year are in prospect, the sheer cost of new infrastructure is impossible to justify.

An example of this is the study undertaken by the New Mexico Transportation Department into the feasibility of an express rail link in the Rio Grande Corridor between Alburquerque and Santa Fe. The original idea was to build a 106 km (66 miles) Shinkansen-type line, but the feasibility study showed a TGV-type approach to be far more cost-effective, involving the building of only 28 km of new track and accepting a speed limit of 145 km/h (90 m.p.h.) on existing track. The difference in journey time between the 'TGV' and 'Shinkansen' approaches would be insufficient to generate enough traffic to cover the large difference in capital costs.

Upgrading existing networks

A similar approach to the French regarding the construction of new lines has been adopted by West Germany. The pre-war Deutsche Reichsbahn had been developed in order to serve the mainly east–west traffic flows across the north of the country and main lines to Berlin. The post-war partitioning of Germany resulted in a totally different pattern of travel flows, with the nation's commercial axis shifting through 90° to become predominantly north–south.

This resulted in lines which had previously seen little traffic becoming major rail arteries. The initial reaction was to upgrade existing track, which in some cases involved major rebuilding operations. As in Britain, speed was not seen as particularly important until the 1960s. The main emphasis was on electrification and, perhaps, significantly, improving standards of comfort. But, as road and air competition grew, so Deutsche Bundesbahn (DB) began to step up its operational speeds from a maximum of 140 km/h (87.5 m.p.h.) to 160 km/h (100 m.p.h.).

The line between Munich and Hamburg had now become a key north–south route, much of which had been rebuilt, with the opportunity taken to include sections capable of 200 km/h operations. On this, DB introduced Europe's first 200 km/h passenger service in May 1966, but within two years the maximum speed had been cut back to 160 km/h. As in France, it was found that conventionally designed locomotives caused heavy track wear. Maintenance costs were 20 per cent higher than at 160 km/h—more than the extra revenue generated by the higher speeds! Greater speed was still desired, but not at this cost.

DB therefore sought—as in Britain, France and other nations—railway vehicle suspension designs that would produce acceptable track forces at higher speeds. For a while, DB contemplated the use of tilting trains, but the continuing need to upgrade a number of lines and to alleviate traffic 'bottlenecks' meant that, as in France, the construction of a number of new lines was economically and politically feasible. In 1970, DB presented to the German Government plans for seven new lines (*Neubaus-trecken*), totalling 950 km (590 route miles), with a top design speed of 300 km/h, together with the upgrading of 1,250 km (775 miles) of key existing lines to 200 km/h (125 m.p.h.). Together, the construction of the new lines and the upgrading of existing routes yielded improvements to the whole of Germany's inter-city network. It was essentially the same stance as was argued by SNCF for the TGV.

The 1973 Federal Transport Route Plan broadly accepted this approach, with authority being given for the construction of the first two *Neubaustrecken*. These are Mannheim–Stuttgart (99 km/61.5 miles) and Hanover–Würzburg (324 km/203.2 miles), construction of which began in 1973 and 1976 respectively (see Figure 29). Both these investments were appraised using cost-benefit analysis which, as Roberts (1984, p. 16) noted, included 'cost changes arising from modal transfer, effects on traffic safety and environmental impact.' For the first two new lines, 'the overall benefit at 1983 prices is predicted at DM 1.5 billion per annum, which exceeds the overall costs by a factor of four.'

But unlike the TGV, which was financed by commercial loans, the German lines are being built using interest-free Federal loans, which very much improves their investment economics. Even so, another crucial aspect is that the design of these new lines has been changed from that of a dedicated fast passenger line to one for mixed passenger and freight traffic. Although building the new lines for heavy freight trains raises construction costs by 10–15 per cent, the investment case required the lines to be used for both passenger and freight. Unlike the TGV, passenger traffic flows alone were insufficient to substantiate the construction of these lines, even under cost-benefit

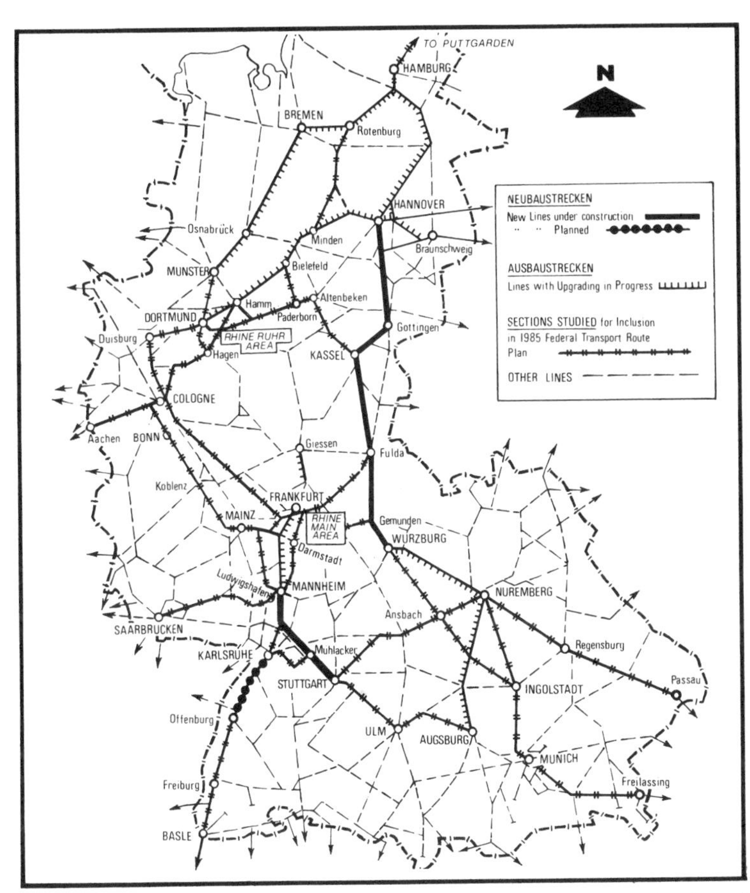

Figure 29. Deutsche Bundesbahn plans for upgraded and new lines
Source: Modern Railways

analysis. This has resulted in the planned operational speed for the passenger trains on the new lines being restricted to 250 km/h, with a design approach to fast trains much akin to the British IC225 project, involving the use of the mixed traffic E120 locomotive. A fixed formation 'Inter-City Experimental' (ICE) train has been built, capable of 350 km/h (218 m.p.h.), but this is purely a technological test-bed and is not a prototype passenger train.

By 1980, authorization had been given for the upgrading of 1,097 km (682 miles) of existing routes to 200 km/h and the 1980 Federal Transport Route Plan identified a further 874 km (543 miles) of routes to receive a similar treatment. Broadly speaking, this is similar to British Rail's upgrading of the 633 km (393 miles) East Coast main line and the 307 km (191 miles) London–South Wales line to 200 km/h. Within these sections of track, 200 km/h capability is not continuous but is created where existing curve alignments (or easily achieved realignments) are feasible. The main difference has been that in Germany the upgrading involved electrification, whereas in Britain it was for diesel HST operations.

DB's progress on such upgrading has been quicker than the construction of the new lines, largely owing to West Germany's complex planning laws (see Freeman Allen, 1985). The two *Neubaustrecken* lines are not expected to be operational until 1991, although some sections will come into use from 1987.

Indeed, although there remain plans for further new lines, the emphasis seems to have shifted towards the upgrading of existing routes. The most likely candidate for another new line is along the Cologne–Frankfurt corridor. Unlike other routes in Germany, this has a traffic potential in the Paris–Lyon TGV league, which would make a dedicated passenger line a viable proposition. However, what is most likely to tip the balance are the plans for the Paris–Brussels–Cologne TGV which, coupled with the Channel Tunnel link to Britain, would open up a vast international traffic potential for any extension of this route beyond Cologne.

Overall, although there are similarities between the German and French approaches to fast train developments, there are some telling and important differences. Both countries have a stable political commitment to the development of their railway system, which is clearly crucial when large, long-term investments are at stake.

Both governments see railway planning at a national level, part of integrated plans for transport developments. The absence of such a viewpoint in Britain and the United States, and its breakdown in Japan, explain a lot of the contrasts between these countries' rail strategies.

Both West Germany and France, for their own unique historical reasons, have faced capacity problems with their existing rail network that has resulted in the construction of new lines being a viable option. Both countries use cost-benefit techniques to assess railway developments. As such, both countries have been able to adopt an approach to fast rail developments that has involved a technically simpler path than in a country such as Britain where investment in new, well-aligned track, is politically and economically ruled out. Were cost-benefit used in Britain, although substantial sections of new track are unlikely simply because no one route suffers from major capacity problems, the major rebuilding of a route such as the West Coast main line could feature.

But it is the differences between the German and French experience that are, perhaps, most telling. Whatever appraisal system is used for an investment, it requires some way of relating the costs of that investment to its benefits. The Germans found that it was only by including freight traffic that they could substantiate their new lines, even given the use of cost-benefit analysis and interest-free Federal loans. For much of their rail network, the upgrading of existing track for 200 km/h operations has been the preferred option. In this, and in DB's approach to mixed traffic rolling-stock, there are in fact greater similarities to British Rail's approach than to that of France.

Wickens (1985) considered that unless a route had the capability to generate twelve million passengers a year, the construction of a new line would not be feasible. The Germans have managed to justify new lines with lower flows than this by including freight traffic. However, although specific factors, such as geography, land costs and the like will affect the investment case for any specific project, the broad principle remains that a threshold must exist below which the 'new build' option ceases to be viable. If the market is not big enough then that scale of investment is simply not worthwhile.

In such circumstances the approach adopted by Britain, and used by Germany for the bulk of its high-speed developments, is therefore necessary. There may be circumstances in which the new-build option is viable but, in general, the upgrading of the existing rail network is likely to be the best investment option available. This is bound to involve greater constraints. In the cases of Germany and France, these constraints were taken to represent the limit to fast train developments, with speeds no greater than 200 km/h envisaged. In the case of Britain, and in some other nations, the constraints of existing track alignments and operational practices have in fact stimulated a more innovative approach. Despite the difficulties experienced by the APT project, the viability of 250 km/h operations on existing curved lines has been proved.

Is there a market for MAGLEV?

If the construction of even a TGV-type new line requires the generation of very high traffic flows to substantiate such an investment then this has a major bearing on even more innovative ground transport systems. As was noted in Chapter 4, the development of the Hovertrain as an Inter-City concept was axed in favour of developing Britain's conventional railways. Yet elsewhere, interest in this technology has been maintained. Both the Japanese and the West Germans have invested large sums in the development of high-speed Magnetic Levitation (MAGLEV) trains. MAGLEV trains offer low track costs and the potential for very high speeds. The Messerschmitt-

Bolkow-Blohm *Transrapid* MAGLEV has run at 500 km/h (313 m.p.h.) on its test track at Emsland, with the Japanese not that far behind.

But despite the fact that MAGLEV offers such a speed potential on a track that is cheaper to build and maintain than for conventional railways, it is hard to envisage the widespread use of MAGLEV for inter-city operations. All the big traffic-generating routes are already served by conventional trains and MAGLEV would therefore have to be able to have a much greater traffic-generating potential than, say, a TGV-type approach, in order to pay for the increased investment costs in carving a new route into existing cities.

The Germans, for example, pressed the French to consider a MAGLEV option for the cross-national TGV Nord route from Paris to Brussels and Cologne. Although the MAGLEV line is cheaper to build *per kilometre*, the inability to run MAGLEV over existing tracks into Paris, Cologne and other cities on the route resulted in track costs 50 per cent *higher* than for the TGV option, where existing lines would be used in these cities. Running at over 500 km/h, the MAGLEV would manage Paris–Brussels in 1 h 6 min, but this would only be twenty-four minutes faster than the TGV. Given that the general order of magnitude of journey time is the same, there was no way in which the MAGLEV would generate sufficient additional traffic to substantiate its greater capital cost. The MAGLEV option was dropped.

MAGLEV would only really begin to generate significantly more traffic than a fast conventional train over longer distances—1,500 km (930 miles) and over—where its higher speed would result in a significantly lower journey time than fast conventional trains. MAGLEV would then have the same kind of time advantage as air travel. But to what extent would MAGLEV be a better investment than the air travel option itself? The financial risks involved in building a 1,500 km (930 miles) MAGLEV would be enormous, whereas its traffic-generating potential would be very unpredictable.

Wickens (1985, p. 512) aptly sums up the situation:

> Whether the Japanese or the Germans will have built a (MAGLEV) route with one of these systems by the end of the century depends on whether the technological push can overcome the dread of the marketing risks—it will be a very complex decision-making process, involving considerations about the exportability of the system, national pride, and other such imponderables.

Given an existing railway network, any totally new system, such as MAGLEV, has to be able to offer a very considerable improvement in either cost or journey times in order to justify the capital investment needed to yield that improvement. This is what happened with air travel over medium- to long-haul routes, resulting in the elimination of the transatlantic liner and America's transcontinental passenger railroads.

For up to 1,500 km, however, improvements to existing railway networks can produce journey times marginally lower than MAGLEV, with around the same traffic-generating potential. And since this primarily involves the upgrading of existing lines, the cost is *lower* than MAGLEV. Above 1,500 km, MAGLEV would generate significantly more traffic than a fast conventional railway, but in terms of cost and journey time it would be no more competitive than air transport. The technology exists, but there is little evidence that there is any real market for it.

At most, MAGLEV is likely to be applied selectively. Given political backing, MAGLEV may be used when a fast, dedicated passenger line is required owing to congestion on existing rail lines or on routes which, for historical reasons, have no existing rail connection. The prime candidate in the latter category is a plan for a Los Angeles to Las Vegas MAGLEV link, involving a 380 km journey. However, its most likely application will be a development of the role fulfilled by the Birmingham MAGLEV: for short city to airport links and other similar functions.

Alternatives to high speed?

Finally, it is worth turning to railway development strategies that do not primarily rely on speed for success. The power that senior mechanical engineers have historically held in Britain's, and other developed nations', railways has tended to result in rail innovations that are predominantly technical. Fast trains are basically an engineering approach to rail innovation. Perhaps the most stark contrast, where innovation has been in the railway *system* rather than the hardware, is supplied by Swedish Railways (SJ).

In the early 1970s, an inquiry into Sweden's railways concluded that the competitive position between road and rail was unjust, owing to the burden of fixed costs that the railways bore in their track infrastructure. It was proposed that, if the state relieved SJ of the cost of this, a large reduction in fares could be achieved. Even though this would actually increase the state subsidy to rail, the increase in passenger traffic would be such that the cost-effectiveness of the subsidy would increase.

This fare-cutting initiative was introduced in the summer of 1979, with a general fares reduction of 30 per cent for long distance journeys. In addition, rebate cards could be purchased for £20, entitling the user to a further reduction of 40 per cent in off-peak periods. A traffic increase of 10 per cent was predicted, insufficient to cover the loss of revenue from the fares reductions which it was thought would result in an overall drop in the scheme's first year of £7 m. By November 1979 it was clear that the effect of such a massive fares reduction had been grossly underestimated. A net *increase* of £9 m in revenue was what actually occurred!

The question as to whether too much emphasis has been placed on speed, and technical innovations in general, is one which deserves careful consideration. Clearly, speed has played a major role in attracting and maintaining rail passenger traffic, but this success has led to other aspects being neglected. The focus has been on technical rather than marketing and organizational innovation.

Marketing studies for British Rail in recent years have revealed that even for the kind of traffic where speed has previously been seen as rail's prime weapon (the business sector), other aspects have been found to be as important. In particular, the quality of the service offered, in terms of ride, seating, environment and on-train services, came out as the passengers' top concern. Consequently, British Rail has undertaken a programme of improving the quality of Inter-City services, involving the refurbishment of existing coaches and, in particular, the launching of a number of top-quality Pullman services on routes to London from northern England. These have successfully boosted first-class business travel by as much as would have been achieved by a 30 km/h (19 m.p.h.) increase in average journey speeds! For example, despite stiff competition on the Leeds/Bradford to London Heathrow route, within a year of its introduction in 1985, the HST Yorkshire Pullman had seen a 20 per cent growth in traffic.

Further studies, commissioned from the Cranfield Institute of Technology (see Abbott, 1986), have examined the effects on passenger revenue of poor reliability of rail services. These suggest that if operational methods and selective investments were made to significantly improve reliability then, in the long term, traffic gains of 10 per cent to the Inter-City sector would result.

These market-based studies, which have focused on the non-technically based influences on rail passenger traffic, by no means deny the passenger-generating effects of speed. But rather than being *the only* way for rail to compete, speed is one of a number of elements, improvements to which are beginning to be sufficiently understood to be costed and compared. This helps to explain BR's now cautious approach to fast train development. Speed is not necessarily the most cost-effective or easiest measure to implement.

There are dangers in swinging too far away from technical innovations, however. Swedish Railways are convinced that the low-price drive cannot alone sustain passenger traffic levels indefinitely. They too are seeking improvements in the quality of rail services, involving electrification, better-quality rolling-stock and, of course, their own tilting 200 km/h train. Although the developed world will see a number of new fast rail projects, there are to be other developments of wider applicability and significance. It seems that an awareness of innovatory hardware, marketing and operating systems, matched by ways of evaluating the cost-effectiveness of each, may well prove to be the really significant characteristic of rail developments through to the next century. They may not have the instant appeal or romance associated with 'technical fix' fast trains, but without them the fast train may never come to fulfil its true potential.

10 The limits of technological innovation

What constrains technology?

What is perhaps the most striking feature of the development of fast passenger trains world-wide is the fact that technological innovation has played a relatively minor part in the eventual outcome. It is true that, initially, the 'hunting' problem did represent a straightforward technical barrier to fast train operations. Even with the largest rail passenger market in the world (and hence an ability to invest the largest sums!), the Japanese Shinkansen, the world's fastest traditionally developed railway, was limited by technical constraints to 210 km/h.

But the 1960s saw the effective elimination of all such technical barriers to fast train development. Today a speed of 300 km/h on well-aligned conventional railway track is standard European practice and 400 km/h seems perfectly feasible. Magnetic Levitation (MAGLEV), though still in its technical infancy, offers the potential for even higher speeds. The German *Transrapid* 06 MAGLEV vehicle has already run at 500 km/h and, technically, there seems little reason why MAGLEV vehicles operating as fast as aircraft could not be in service within the next ten to fifteen years.

The technology has ceased to represent any real barrier to innovation. But the rapid elimination of technical barriers to fast trains has opened up a large, overlapping mesh of interrelated historical, economic, political, social, geographic and organizational factors which have now become crucial to the form and extent to which innovation in fast ground transport is practical. These factors have always been present, but in the past have been muted because the scope for fast trains has been so technically limited. The particular mix, and liability to change, of these factors goes a long way to explaining contrasts and divergencies in fast train developments between nations. Indeed, it is at the extremes of these constraints that the most innovative approaches are to be found: firstly, where there are relatively few constraints and where the potential for technological innovation can be fully explored; secondly, where the constraints are so great that they in themselves stimulate a high level of innovation. The two extremes of stimuli for innovation may well be represented by MAGLEV at one end and by the APT at the other, with less innovative developments, such as the TGV, Shinkansen and ICE, falling in between.

Economic constraints

In simple economic terms, the amount that it is viable to invest in a project must be related to market potential. This is true whatever form of investment appraisal is used, be it strictly commercial or a systems-based cost-benefit approach. Only where the market potential is very large is it worth contemplating ground transport systems that require new network infrastructure. Although the level of traffic at which this becomes a possibility will vary according to the investment appraisal method, local land values, geography, etc., it seems unlikely that a major new ground transport system such as MAGLEV will really be viable unless it can attract passenger flows of the order of fifty million per annum.

Demand for fast transport is a characteristic of the developed world, and it is in the developed world that the most extensive networks of existing railways, roads and air services are to be found. Any innovative system that requires new infrastructure must have a massive advantage over existing transport systems, either in terms of cost or performance, to have any chance of success. The basic characteristics of MAGLEV may make it competitive in a 'new build' situation, but it does not have the sort of advantage that makes it a serious contender for replacing existing transport methods. Its technology may be advanced, but its economics and performance are of the same order as rail and air, so further improvements to these existing systems will always be a more cost-effective use of investment funds.

Studies in Europe and the United States have shown this to be so, with the development of conventional railways judged, by both private and public assessment methods, to be far more cost-effective than alternative fast ground transport systems. There may be a limited number of locations throughout the world where, because of specific circumstances, a new MAGLEV link is viable, but these are very few.

This illustrates one major constraint upon technological innovation. The existence of a well-developed infrastructure or product support system makes the entry of a rival based on a different infrastructure difficult or impossible, even though the technology of that rival may be superior to the existing product. This may simply put off the eventual entry of the new product, while it is developed to a more advanced level; may restrict it to a limited market (as seems to be the case with MAGLEV); or may act as a virtual total barrier, as was the case with the Brennan monorail at the turn of the century.

Railways are but one example of this 'established product and infrastructure support system' effect on rival innovations. As was noted in Chapter 4, petrol-engined technology has a similar advantage. Another, smaller-scale, example is the difficulty that compact disc manufacturers have experienced in establishing their technically superior product given that people already have an 'infrastructure' of LP decks and existing stocks of LPs. Any attempt to replace tape cassettes would face a similar problem. It takes a real breakthrough to seriously threaten any widely accepted technology.

Under such circumstances, fast ground transport innovations have focused on improvements to the basic 'steel wheel on steel rail' technology. Nevertheless, within this there have been differing levels of technological innovation which have, at their root, economic constraints. Where the traffic-generating potential is high, major infrastructure improvements represent a viable approach, as has been the case with the TGV, in Germany, Italy and some other nations. Under these circumstances, relatively modest improvements to existing rail technology are all that is necessary to achieve high speeds and to keep railways competitive.

But at lower levels of traffic-generating potential, the case for investment in upgraded infrastructure

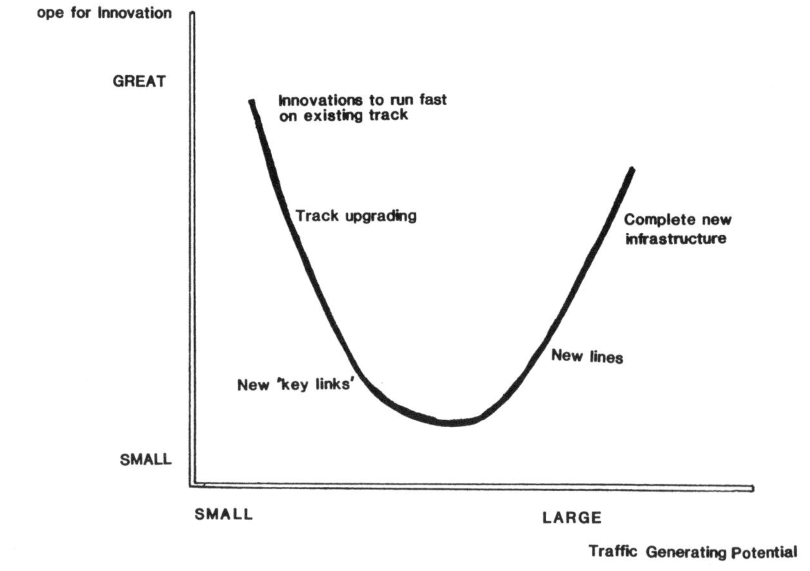

Figure 30. Diagrammatic representation of the relationship between market size and level of innovation

weakens. The emphasis shifts from infrastructure to the vehicles that run on that infrastructure. In terms of the railways, once traffic-generating potential drops below ten million passengers per annum, the new-build option becomes very hard to substantiate. At even lower levels, even upgrading existing track becomes of dubious value. This additional constraint therefore results in the requirement for a higher level of innovation in the fast rail vehicle itself. This requires more advanced suspension designs, body tilt (if routes are curved and speeds over 200 km/h are contemplated), compatability with mixed traffic operations, systems that fit in with existing signalling and other operational methods, plus a host of other detailed factors.

Figure 30 expresses the broad economic stimulus/constraint on rail innovation in a diagrammatic form. At both low and high levels of traffic-generating potential the need for innovation is greatest, but the form it takes is very different.

National differences

In so far as a general model of rail innovation is possible, the economic concepts expressed in Figure 29 are almost as far as it is possible to go. Beyond this, the form which fast train innovations take seems to be very much determined by specific forces and constraints, which in themselves can act as stimuli to innovation. In some cases these forces and constraints are a product of social and political factors peculiar to one country (the low value assigned to rail developments in Britain compared to elsewhere in Europe, the power of the road lobby, the historical legacy of unplanned rail development), in others, the forces are more general (the effect of the exact form of existing infrastructure on innovative proposals, population distribution, etc.).

Facing economic, historical and political constraints greater than in most other developed nations, British Rail's APT concept was an attempt to develop a high-performance train within these constraints by a radical jump in technology. But the really radical element of the APT was less in the technology of the train than in the engineering approach adopted.

Because of the legacy of their infrastructure and the long life of railway-stock, railway operations world-wide are essentially evolutionary in that only a small element of the whole system can ever be changed at any one time. This evolutionary tradition of rail operations had led to a similar 'cut and try' evolutionary approach to rail engineering. The APT concept endeavoured to separate the evolutionary nature of the operational railway from the then evolutionary research and development process. So, while accepting the operating constraints of the railway business and the evolutionary approach that this required, BR Research adopted a radical 'scientific' approach to design.

It was accepted that the *operational* side of the railway business had to innovate in an evolutionary manner. But why should this determine the approach to engineering? A radical approach to research and development should be perfectly compatible with the need to apply that research and development in an evolutionary manner.

The example of overcoming the 'hunting' problem illustrates this. The constraints of railway operations were taken for granted. Any new development would have to be gradually introduced and be compatible with operating with other trains using the old design of bogies. But this operational evolutionary approach was unrelated to the scientific approach adopted to raise critical speeds and eliminate hunting. So long as the operational context in which innovations are applied is understood, a radical approach to research and development can be compatible with an inherently evolutionary industry such as rail.

Managerial constraints

The design concept of the APT clearly embodied this understanding of the evolutionary nature of railway operations and a clear understanding of the commercial and operational environment of British Rail. But, although the economic constraints were largely understood, the managerial constraints to innovation were hardly appreciated. The main problems faced were simply ones of managing the development process. The 'research team' approach, although well suited to a design job in a research department, was too inflexible for a development task. The conventional method of development in BR, that of a job being seen through the functionally organized M & EE Department by the head of the division, was only really capable of dealing with evolutionary innovations. Neither method was really suitable for the APT and the management method adopted, with the APT Project Group being in (but not really part of) M & EE, was crude compared to the technology it was intended to manage.

This shows how difficult it can be for an industry whose whole method of organizing and managing itself is geared to a particular innovation approach to try a different way of innovating. Established procedures, methods of allocating staff and organizing departments need to be changed. This is bound to cause personality problems as individuals see their position threatened, as well as internal rivalries between new and old groupings. The upheaval this causes, or the fear of it, may well prevent any innovation from being developed, even though it may be technologically and commercially viable. Even if the innovation goes ahead, it may simply get bogged down and die because the existing management structures cannot cope. Before embarking on an important innovation, a company's management should not only ask if they have the technical expertise to see the project through, but whether their management structure can cope with it.

In the case of the APT (and also of the Metroliners in the United States), it may well have looked as if problems were technological in nature, as particular 'bugs' failed to be cured. But this was far from the truth. It was more a matter of appropriate organization and project management. It was only when the APT project organization was restructured along 'matrix' lines with an experienced project manager in charge that the technical bugs were finally eliminated. The lack of major developmental problems with the TGV owed a lot to its 'matrix' project management. It was not just that the TGV was a technically easier project to manage.

This raises the question as to how many innovation problems are really technological or whether they are a matter of the way in which a project is organized and managed. Have other 'technical failures' in the past really been project management failures? An awareness that you can only innovate as fast as the organization can manage innovation is one that can come too late.

It is remarkable to note how close the APT got to success and how close the HST was to failure. If there had been a longer trail period on the HST prototype and the problems with its engines recognized, it is likely that the production run would have been held up. In practice it took longer to cure the HST's problems than those of the APT, and some problems are still persisting. As it is, with a quickly built fleet of HSTs, the bugs are having to be sorted out in service. This, indeed, was what happened in the case of AMTRAK's Metroliners. If a fleet of ten APT-Ps had been built, as was originally proposed, the same approach could well have been used, with high maintenance used to buy reliability. Or if the reorganization of M & EE were to have come a couple of years later, this book might well have featured the successful APT and the unreliable, soon to be scrapped, HSTs! It is remarkable how close the boundaries between 'success' and 'failure' are.

British Rail's reorganized M & EE Department can now

adequately manage innovations that the old organization was unable to cope with. It is a very different department now to what it was in the mid-1970s. But the APT fell victim to being both an innovation that needed the new management structure before that structure existed and to the general upheaval that creating that management structure produced. By the time all this was over and had settled down, market conditions had changed such that there was no longer a commercial need for the APT.

'All or nothing' or 'Softly—Softly'?

Following the APT, BR has adopted a cautious approach to innovation, with a concentration on 'core' innovations needed for the IC225 and the breaking up of the project into stages. The development of the high-performance locomotive and basic coach design, operating at 200 km/h, will come first, followed (possibly) by the tilting coaches, and then moving on to 225 km/h. The strategy is more evolutionary, but the radical scientific approach to design and development is as strong as ever. What has happened is that BR has learnt how to apply this innovative scientific approach to its necessarily evolutionarily-geared operations.

From this must arise the question as to why such a large leap in technology and operating performance was attempted by the APT. The innovations could have been gradually introduced over a series of train designs which would have considerably reduced the risks involved. Some developments require an 'all or nothing' approach in order to succeed and win a large new market. This is not generally the case with railways, or with the majority of other industries either. In actual practice a gradual introduction of APT innovations is what has happened; the high-speed bogie/suspension design has been introduced on a series of trains already in service; so has streamlining and methods of reducing unsprung mass; the final drive of the Class 91 Electra has been developed from export locomotives built by GEC and, as noted above, the coaches will be produced first in a non-tilting and then in a tilting version.

Product design and marketing

The reorganization of product development within British Rail in the wake of the APT project has produced more than a reformed project management. It has been part of a long-term reform of the structure of the railways designed to provide targets and objectives for the main business sectors and structures that allow different alternatives to be evaluated and consequences followed through. Basically it represents moving towards a 'systems' approach to management and some of the aspects of the IC225 project illustrate this systems approach well.

A greater awareness of marginal design assessment has developed. Increasing technical complexity often involves key thresholds, whereas the marketing value of the product does not. For fast trains, such thresholds are associated with speed: new braking, new signalling, use of tilt/realigned track, etc. The InterCity 225, by using a design speed just below a series of these thresholds, eliminates the need for a number of innovations that were incorporated into the APT, whose design speed was just above them. As subsequent studies showed, the market impact of the higher technology would have been minimal.

This brings us back to the discussion begun in Chapter 4 on the initial stimulus for innovation. Characteristically, proposals for rail innovations have originated from railway engineers and not from the operational or management side of the railway. There are benefits and disadvantages in both 'technology push' and 'market pull' and one objective of BR's reorganization in the last decade has been to try to establish a balance between these factors in recognition of the fact that 'technology push' has, in the past, had the upper hand. In Chapter 6, the commercial impact of the HST was found to be less dependent on speed than was expected and in Chapter 9 the impact of marketing and pricing innovations was contrasted with that of technological innovations such as fast trains.

Although speed is still seen by British Rail as an important factor in generating rail passenger traffic, the simple speed/passenger numbers correlation has been replaced by wider based studies of what influences rail passenger traffic. For different rail markets speed, cost, comfort, convenience, access to stations, and 'image' all have a role to play. This view is not universally shared, with many railways still wedded to the simplistic speed = passengers formula.

The widening of BR's approach to winning passenger traffic from a simple 'technical fix solution' of fast trains to a programme of identifying key decision-making features and then applying a mix of technical, marketing and organizational innovations to influence these factors explains the reduced need for a large fleet of fast trains. It also raises the question of whether there has been too great a reliance on simple 'technical fix' innovations rather than innovations in marketing, organization and innovation management. Have we depended too much on engineers presenting technical-fix solutions to operational management, leaving a lack of real planning over the use of different types of innovations?

This raises questions that range far beyond the bounds of this case study. The Finniston Report ('Engineering, Our Future', HMSO, 1980 *see* Committee of Enquiry into the Engineering Profession), which investigated the organization of the engineering industry as a whole, was highly critical of the lack of planning in the British engineering industry compared to Germany and Japan. In particular the report considered that British engineering companies were handicapped by a 'lack of adequate engineering input to marketing business planning activities and by a lack of sufficient market input to engineering activities.'

The accusation is that designers and engineers wield sufficient power to impose technically-led (and often inappropriate) innovations upon an industry. They may seem successful and there may be effective commercial input once a technical innovation is proposed, but the whole technology-led approach discourages alternative, non-technically innovative, options. It has taken a long time for British Rail's marketing and management to get their marketing and organizational innovations evaluated as seriously as the 'technical fix'.

Clearly, the best kind of organizational structure is one which can respond positively to either type of innovative initiative; where marketing is able to effectively evaluate technical proposals and where engineering has a clear understanding of marketing possibilities in order to innovate where it is most needed. Complementary technical, organizational and marketing innovations then become possible. Probably the most distinctive feature of the InterCity 225 is that its design in fact reflects this multiplicity of innovative approaches: the high-speed technology, the well-designed, market-oriented Mk4 coaches, the robust adaptability to changing circumstances and the systems-based use to which locomotives and coaches will be put.

This book began by looking at a 'might have been'—the technologically spectacular gyroscopically stabilized monorail; an innovation that got nowhere. Clearly, technology is constrained, but to portray rail technology as reaching a limit would be misleading. The picture that emerges from this study is one of an intricate mixture of constraints and stimuli: a mesh of different levels and often quite specific factors that, rather than limiting technology, determine the direction in which technological developments move.

The mesh of economic, political, social, historical and managerial factors that has had such an influence upon fast train developments is present, in different combinations and to differing degrees, in all parts of society. Recognizing them and the way in which they constrain and stimulate innovation is as much a part of successful innovation as the research and development work on the innovation itself. Indeed, it is a prerequisite to any truly successful innovation.

References

Abbott, James, 1986. 'Serving the Customer', *Modern Railways*. January, pp. 27–8.

ACARD, 1980. *R & D for Public Purchasing* (report of the Advisory Council for Applied Research and Development). London, HMSO.

Aldcroft, Derek H. 1969. 'Innovation on the Railways: the lag of diesel and electric traction', *Journal of Transport Economics and Policy*. January, pp. 96–107.

Anon, 1973. 'APT Programme Moves Ahead after Searching Reappraisal', *Railway Gazette International*, December, pp. 469–71.

——, 1977. 'Prototype APTs take shape at Derby', *Railway Gazette International*. January, pp. 27–31.

Bennett, R. C. & Cooper, R. G., 1979. 'Beyond the Marketing Concept', *Business Horizons*, June 1979, pp. 76–83.

Boocock, D. 1985. 'Future Electric Locomotives and Coaches for the InterCity Business', *Railway Gazette International*, June.

—— & Newman, M. 1976. 'The Advanced Passenger Train', *Proceedings of the Institution of Mechanical Engineers, (Railway Division)*. **190**, 62/76, pp. 653–63.

—— & King, B. L. 1982. 'The Development of the Prototype Advanced Passenger Train', *Proceedings of the Institute of Mechanical Engineers*, **196**, pp. 35–46 & S21–S34.

British Railways Board. 1963. *The Reshaping of British Railways* (The Beeching Report). London, HMSO.

—— 1984. *InterCity—Into Profit*. British Railways Board, December.

British Transport Commission, 1956. *Proposals for the Railways*, Cmnd 9880, London HMSO.

Campbell, I. 1980. 'Railway Gazette puts the APT in perspective', *Railway Gazette* supplement, May.

Committee of Inquiry into the Engineering Profession. 1980. *Engineering our Future* (The Finniston Report). London, HMSO.

Cottrell, Richard 1981. 'Sweden's Cut Price Rail Fares Gamble Pays Off', *Modern Railways*. October, pp. 457–9.

Evans, Andrew W. 1969. 'Intercity Travel and the London Midland Electrification', *Journal of Transport Economics and Policy*. January, pp. 69–95.

Faith, Nicholas. 1985. 'The World and its Railways: 4—The Fast Track', *The Economist*, 14 September, pp. 23–9.

Ford, Roger. 1984. 'Engineers Eclipsed', *Modern Railways*. August, pp. 413–17.

——. 1985. 'Business Sectors Challenge BR's Engineers', *Modern Railways*. January, pp. 21–5.

—— 1986. 'Electra', *Modern Railways*. April, pp. 201–3.

Freeman, Christopher. 1982. *The Economics of Industrial Innovation*. 2nd edn, London, Frances Pinter.

Freeman Allen, G. 1978. *The Fastest Trains in the World*. London, Ian Allan.

——, 1985. 'High Speed Lines Open in 1991', *Modern Railways*. June, pp. 303–8.

Grange, Kenneth. 1983. 'Designs for British Rail', in S. Bailey & J. Ward (eds), *Kenneth Grange at the Boilerhouse: an exhibition of British product design*. London. Couran Foundation, pp. 46–8.

Hamer, Mick. 1982. 'The Abandoned Passenger Train?', *New Scientist*, 13 May, pp. 406–7.

Hamilton, Kerry & Potter, Stephen. 1985. *Losing Track*. London, Routledge & Kegan Paul.

Haresnape, B. 1978. 'Journey by Design 1948–1973', *Modern Railways*. November, supplement.

Johnson, John & Long, Robert A. 1981. *British Railways Engineering 1948–80*. London, Mechanical Engineering Publications Ltd.

Jones, Sydney, 1973, 'High Speed Railway Running with special reference to the Advanced Passenger Train', *Chartered Institute of Transport Journal*. January, pp. 49–61.

Laithwaite, E. R. & Wilson, J. 1978. *The Rise and Fall of the Tracked Hovercraft*. CEI Committee on Creativity and Innovation, Working Party Case Study 78/01, Institute of Mechanical Engineers.

Langdon, Richard & Rothwell, Roy. 1985. *Design and Innovation: Policy and Management*. London, Frances Pinter.

Lawrence, R. L. E. 1977. 'The High Speed Train', *The Chartered Institute of Transport Journal*. **37**, pp. 260–2.

Ledsome, Colin. 1981. 'APT', *Engineering*. February, pp. 96–106.

Levitt, T. 1960. 'Marketing Myopia', *Harvard Business Review*, July/August, pp. 45–56.

Lorenz, Christopher. 1983. 'Market Research: a fear of feedback', *Design*. December, pp. 31–41.

Modern Railways. XPT–New South Wales' political train, October 1983, pp. 544–6.
Mowery, David & Rosenberg, Nathan. 1985. Government Policy, Technical Change and Industrial Structure: the US and Japanese Commercial Aircraft Industries, 1945–1983', in Richard Langdon & Roy Rothwell, *Design and Innovation: Policy and Management*. London, Frances Pinter, pp. 72–102.
Newman, M. 1973. 'The Research and Development Programme of the Advanced Passenger Train', *Proceedings of the Institute of Civil Engineers*. vol. 55, pt 2, June, pp. 319–33.
Nock, O. S. 1983. *Two Miles a Minute*. Wellingborough, Patrick Stephens.
Oakley, Mark. 1984. *Managing Product Design*. London, Weidenfeld & Nicholson.
Ogilvie, John & Johnson, Brian. 1975. 'Railways: accounting for disaster', *New Scientist*, 23 October, pp. 228–9.
Perren, Brian. 1985. 'The TGV Atlantic Project', *Modern Railways*. November, pp. 576–80.
Potter, Stephen & Roy, Robin. 1986. 'Research and Development: British Rail's Fast trains', Block 3 of *T362 Design and Innovation*. Milton Keynes, Open University Press.
Roberts, John & Woolmer, T. 1984. *BR: A European Railway*. London, Transport and Environment Studies.
Rothwell, R., Schott, K. & Gardiner, J. P. 1985. *Design and the Economy: the role of design and innovation in the prosperity of industrial companies*. 3rd edn, London, Design Council.
Roy, Robin & Wield, David (eds). 1986. *Product Design and Technological Innovation*. Milton Keynes and Philadelphia, Open University Press.
Sephton, B. G. 1974. 'The High Speed Train', *Railway Engineering Journal*. **3**, September, pp. 22–41.
Shilton, David. 1982. 'Modelling the Demand for High Speed Train Services', *Journal of the Operational Research Society*. **33**, pp. 713–22.
Takayama, Akira, 1982. 'Shinkansen Network Unrolls on Honshu', *Railway Gazette International*. October, pp. 813–15.
Taylor, K. 1979. 'Interphase', *Proceedings of the Institute of Mechanical Engineers*. **193**, December, pp. 369–71.
Transport and Environment Studies (TEST). 1985. *The Company Car Factor*. TEST, January, London.
Twiss, B. C. 1980. *Managing Technological Innovation*. 2nd edn, Harmondsworth, Longmans.
Tyler, Paul. 1984. *The Advanced Passenger Train: What?, Where-*

fore?, and Where now?. Unpublished BSc Project Report, April, Middlesex Polytechnic.

Walgrave, M. Michel. 1985. *The Paris–Southeast TGV System*. SNCF Planning and Research Department Paper, SNCF Paris, September.

Wickens, A. H. 1971. 'The Advanced Passenger Train', *Advancement of Science*. March, pp. 208–15.

—— 1973. 'Stability of High Speed Trains', *Physics in Technology*. **4**, No. 1, pp. 1–17.

—— 1977. 'Dynamics and the Advanced Passenger Train', in *Speaking of Science, Proceedings of the Royal Institution*, **50**, pp. 33–66.

—— 1983. 'Research and Development on High Speed Railways—Achievements and Prospects', *Transport Reviews*, **3**, No. 1, pp. 77–112.

—— 1985. 'Railway Technology in 2001—promise and reality'. Paper delivered to the Chartered Institute of Transport, reproduced in *Modern Railways*. October, pp. 509–13.

Williams, Hugh. 1985. *APT—A Promise Unfulfilled*. London, Ian Allen.

Index